Explaining the Computational Mind

Explaining the Computational Mind

Marcin Milkowski

The MIT Press
Cambridge, Massachusetts
London, England

© 2013 Massachusetts Institute of Technology

All rights reserved. No part of this book may be reproduced in any form by any electronic or mechanical means (including photocopying, recording, or information storage and retrieval) without permission in writing from the publisher.

MIT Press books may be purchased at special quantity discounts for business or sales promotional use. For information, please email special_sales@mitpress.mit.edu or write to Special Sales Department, The MIT Press, 55 Hayward Street, Cambridge, MA 02142.

This book was set in Stone Sans by Toppan Best-set Premedia Limited. Printed and bound in the United States of America.

Library of Congress Cataloging-in-Publication Data is available.

ISBN 978-0-262-01886-9

10   9   8   7   6   5   4   3   2   1

Contents

Preface

This book is about explaining cognitive processes by appeal to computation. The mind can be explained computationally because it *is* computational; this is true whether it is engaging in mental arithmetic, parsing natural language, or processing the auditory signals we attend to in order to experience music. All these capacities arise from complex information-processing operations of the mind. My central claim reflects my adherence to realism: a computational account of the mind can constitute a genuine explanation *only insofar as the mind is itself computational*. This stands in stark contrast to computational models of, say, the weather, which, rather than construing clouds and rainfall as the processing of information, describe the purely physical processes underlying the meteorological phenomena.

Why write yet another book about computation and cognition? So much ink has been spilled; positions have been refined and hardened; it's difficult to envisage anything but a scholastic exercise in adding even more distinctions to the discussion. Well, in fact, the reason for writing this book was my growing impatience with extant accounts of computational explanation of cognition, which proved either too limited in scope or too sketchy. There simply wasn't a book that I could refer to in discussions with cognitive scientists and philosophers alike, even if parts of the puzzle were easily available; so, I had to write that book myself.

The term "computation" in cognitive science had been misappropriated to refer to a class of processes that were supposed to occur only during rule-governed discrete symbol manipulation. But this way of framing computation is highly idiosyncratic and leads to numerous misunderstandings—especially because it's hard to say what the term "symbol" stands for—in an interdisciplinary debate taking place between people of varied backgrounds. It is unclear whether, and if so how, these rule-governed processes should differ from rule-following processes or, for that matter, rule-compatible processes. Symbolic computations in cognitive science are supposed to be radically different from operations of classical connectionist artificial neural networks (ANNs). To a software engineer, however, ANNs are just another kind of machine learning. Why would they be noncomputational if we build dedicated computers composed of ANNs? Even worse, why exactly should we refrain from calling neuro*computational* models computational? Is this a kind of institutionalized terminological schizophrenia?

Superficial distinctions obscure a common pattern present in the different branches of cognitive modeling. This, in turn, leads to debates over the alleged supremacy of high-level theories over neurocomputational modeling, or vice versa, and to dismissing whole bodies of research as inessential in cases where it would be more useful to strive for integration rather than elimination.

I suggest that we will be happier if, instead of adding more distinctions, we have fewer of them. Computation—understood generally as information processing—is a basic ingredient in the majority of cognitive theories and models. Granted, the notion is broad, but the alternative ways of carving it are not viable when it comes to saying when a computation is being implemented. The symbolic, semantic, syntactic, and modeling accounts of implementation are plagued by simple yet devastating objections; for instance, one can easily gerrymander the physical processes or entities to which the computations are said to correspond. Real progress in determining when a computational process is being implemented occurs once we have been able to replace the wildcard term "symbol" with something more physically tangible, and the structure of the process is no longer open to gerrymandering. Given that we already have fairly robust theories of causation in the philosophy of science, I suggest that the best way to accomplish these aims is by relying on the causal structure of the world.

Of course, there is another way. Adding a number of provisos to the traditional, semantic, syntactic, or modeling account could block gerrymandering. But that would lead to an explosion of epicycles, and, at any rate, the provisos would boil down to a redescription, in semantic or syntactic terms, of the conditions that differentiate genuine causal structures capable of information processing from mere hodgepodge mixtures of physical states. There is nothing to be gained by such redescription except more verbosity and unnecessary complexity.

Although the vision I offer in this book is much more ecumenical than the received view, this does not mean that I cannot distinguish between the different approaches in cognitive science (or anywhere else the computational idiom is used). Those differences are not really essential, however, when it comes to general methodological principles. If you want to confirm that your computational model is empirically sound, you are not compelled to rely *only* on the specific methodology of Bayesian, biorobotic, cybernetic, connectionist, dynamic, neurocomputational, or symbolic modeling—whatever the case may be. The basic principles of explanation and confirmation are the same for all the methodologies, and this makes comparisons possible even if the mechanisms posited by different methodologies are dissimilar in important ways.

A word about mechanisms is in order. I started to think of physical computation about a dozen years ago. In my doctoral dissertation on Daniel Dennett's philosophy of mind (unfortunately for the English-speaking readers, I wrote it in Polish), I discussed his claim that evolution is algorithmic. The notions I used to describe algorithms and computation coalesced around causation, structures, and processes rather than merely around function terms. Only later did I find that the terminology of organized entities that I had used was in fact equivalent to the vocabulary of the neomechanistic philosophy of science. Having discovered, like Mr. Jourdain, that I had been talking prose, I decided to get rid of my own terminology and settled for mechanistic talk. This is beneficial because the mechanistic account of explanation fits the practice of cognitive science much better than do the competing accounts. As I will show, it fares better than classical functionalism in particular.

Computationalism is here to stay—but it's not what most people had taken it to be. In particular, it does not rely on a Cartesian gulf between either software and hardware or mind and brain. The computational method of describing the ways information is processed is usually abstract—but cognition is possible only when computation is realized physically, and the physical realization is not the same thing as its description. The mechanistic construal of computation allows me to show that no *purely* computational explanation of a physical process will ever be complete. This is because we also need to account for how the computation is physically implemented, and in explaining this, we cannot simply appeal to computation itself. In addition, we need to know how the computational mechanism is embedded in the environment, which, again, is not a purely computational matter. For this reason, computationalism is plausible only if you also accept explanatory pluralism: the proposition that there are acceptable causal explanations that are not spelled out in terms of any computational idiom. This is perfectly in line with the mechanistic philosophy of science.

In chapter 1, I describe four wildly dissimilar models of cognition that are in an important sense computational. The rest of the book defends a vision of computational explanation that makes sense of the research reported in the first chapter. For this purpose, I clarify, in chapter 2, what it is for a computation to be implemented, and I defend a broadly mechanistic theory of implementation. I also tackle a number of skeptical objections here, which is why the chapter is so lengthy. In chapter 3, I argue that, against the background of other philosophical theories of explanation, the mechanistic theory of explanation is the most adequate—both normatively and descriptively. I also discuss empirical validation of computational models there, though the topic of the testing of psychological theories deserves a separate book-length treatment. But we need to know at least the basic principles of testing to see that computational models really are descriptive and explanatory. The task of chapter 4 is to vindicate a notion of representation that is compatible with my mechanistic account of computation and explanation as well as adequate vis-à-vis an analysis of the four cases introduced in chapter 1. I defend a fairly nonstandard account of representation framed in terms of representational mechanisms, which are—at the very least—capable of detecting representational error on their own. Instead of claiming that there is no computation without representation, I say that the slogan is true when you have it backward: no representation without computation. Again, the topic of representation is merely touched upon. Finally, in chapter 5, I show the limits of computational explanation and argue for explanatory pluralism in cognitive science.

Acknowledgments

I am indebted to a number of people for preparing the material for this book. Over the years, I discussed the issues of computation, computational modeling, mechanisms, and particular models with many people, including Darren Abrahamsen, Ron Chrisley, John Collier, Cristian Cocos, Markus Eronen, Antonio Gomila, Lilia Gurova, Edouard Machery, Gualtiero Piccinini, Steven Pinker, Robert Poczobut, Tom Polger, and Matthias Scheutz. Also, participants of numerous Kazimierz Naturalist Workshops (KNEW) were very helpful in developing the views presented here. KNEW is co-organized by Konrad Talmont-Kaminski, whose friendly advice and constant encouragement to publish in English were very important to me.

Mark Bickhard, Paco Calvo, Chris Eliasmith, Paweł Grabarczyk, Vincent Muller, Oron Shagrir, Aaron Sloman, and Barbara Webb read some parts of the manuscript and contributed helpful comments. My students at the University of Warsaw who were exposed to the early version of this book were helpful in pointing out gaps and weak spots in my arguments.

Witold Hensel deserves a special mention: his advice and numerous corrections helped me a lot, in improving both the English and the line of the argument. I also need to thank my beloved Krystyna for bearing with me when I spent numerous evenings working on the book and for frequently disagreeing with my arguments. Alas, nobody can be held responsible for this book's errors, falsities, and mistakes but me.

This work was supported by Polish Ministry of Science and Higher Education grant N N101 138039 under the contract 1380/B/H03/2010/39.

1 Computation in Cognitive Science: Four Case Studies and a Funeral

It used to be a commonplace that cognitive science aimed to account for cognition exclusively by appeal to computer models. Recently though, the situation has changed—as many researchers in the field not only construct dynamical systems, but also reject the traditional computational theory of mind altogether, what once passed for a platitude is now being denied. As a result, computationalism has been pronounced dead on more than one occasion.

Taking a step back from the heated debates that inevitably arise as soon as a new view on cognition appears on the market—often hailed by its proponents as offering novel or even revolutionary insights into the nature of mind—one cannot help observing that, despite prima facie competition from a variety of nonclassical accounts, computation still plays an important part in contemporary cognitive science. Over 80 percent of articles in theoretical cognitive science journals focus on computational modeling (Busemeyer and Diederich 2010). One reason for this is that it is almost a definitional feature of cognitive systems that their behavior is driven by information, and, apparently, no one has the slightest idea how to do without computational explanation where an account of information processing is concerned. Therefore, practitioners of the philosophy of cognitive science would be well advised to try to understand how computational explanation works; computational explanation is precisely what I propose to account for in this book.

1 The Dynamical Approach: What Might Cognition Be, If Not Information Processing?

One recent attack on the idea that cognition must involve computation has been launched by advocates of the dynamical approach; they point out that "rather than computation, cognitive processes may be state-space evolution" within dynamical systems (van Gelder 1995, 346). Cartesian preconceptions notwithstanding, cognitive phenomena are essentially no different from weather patterns or planetary motions, the dynamicists say, and should be explained in terms of state vectors and differential equations rather than, say, Turing machines. In particular, theories in cognitive science need not posit any language-like medium of internal representations over which computations, or rule-governed manipulations, are performed (cf. the characterization of "computation" in van Gelder 1995, 345).

The notion of computation I will defend in this book is much broader— it is not restricted to the ideas of rules and representations presupposed by the language of thought hypothesis. The core of the computational theory of mind, as I see it, is not the truth-preserving transformation of symbols; it is information processing, and information need not be digital or represented in a language-like fashion. Even digital computers are dynamical systems: they can be described mathematically as the evolution of parameters in a multidimensional space. They are not the same as abstract, formal mathematical entities, such as the Turing machine; their implementations are spatiotemporal and physical mechanisms. And when they control anything, control-theoretic formalism, which is mathematically the same as the dynamical approach, seems to be a good tool to describe them. Since the very beginning of computational modeling of the mind, there have been two traditions: (1) cybernetic-dynamical and (2) logical-symbolic (Boden 2008). Some hybrids of both are also possible (e.g., Giunti [1997] uses dynamical formalism to investigate computability).

Can this broader computationalist claim still be denied? Of course it can. It is an empirical bet. If you think that the core of cognition is adaptive behavior, and that adaptive behavior is best explained not by recourse to information but instead to certain kinds of self-organization, you will probably reject it. For example, some writers hold that adaptive behavior is possible only in certain kinds of autonomous systems (Barandiaran and Moreno 2008) or dissipative dynamical structures far removed from thermodynamic equilibrium. The flexibility of behavior is to be explained as self-maintenance of the far-from equilibrium state. In a less physicalist version, this kind of autonomy, called autopoiesis (Maturana and Varela 1980), is to be understood as a kind of cyclical self-organization.

Interesting as these ideas of autonomy are, they do not seem to be specific enough to capture the phenomenon of cognition, which is a narrower category than adaptive behavior. For instance, while describing the behavior of a slime mold as adaptive seems both natural and justified, ascribing cognition to it would strike one as a bit far-fetched. This leads naturally to another important question: what is cognition? I doubt that the matter that can be settled simply by pointing to a definition; this is because definitions of scientific concepts come as parts of theoretical frameworks, and different theoretical frameworks present different visions of what cognition is.[1] The answer of traditional cognitive science is that cognition is information processing in the physical symbol system (Newell 1980) or that it is computation over representations (Fodor 1975). Another important answer is related to the program of modeling minimally cognitive phenomena (Beer 1996; Barandiaran and Moreno 2006; van Duijn, Keijzer, and Franken 2006). The anthropocentric perspective of cognitive psychology is rejected there in favor of a more biological approach. In searching for minimally cognitive phenomena, various authors argue that even bacteria, such as Escherichia Coli (van Duijn, Keijzer, and Franken 2006), and plants (Calvo and Keijzer 2009) are not merely reactive and that their behavior goes beyond hardwired reflexes.

Hence, some writers seem to believe that cognition is equivalent to adaptive behavior. But even if we extend the notion of cognition in this way, the question still remains whether it can be understood in terms of energetic autonomy *only*. Interestingly, Barandiaran and Moreno seem to deny this; they admit that the neural domain is "properly informational" (Barandiaran and Moreno 2008, 336). They go on to explain that they take "information" to stand for the "propagation of dynamic variability as measured by information theory" (ibid., 342). Defenders of bacterial cognition appear to share this conviction:

At least at the level of minimal cognition, it is clear that a thorough understanding of bacterial behavior formed exclusively in computational terms would be incomplete. The characteristics of the embodiment of the E. coli bacterium can teach us about the biological preconditions for minimal cognition. For example, E. coli's rod-like shape diminishes the impact of Brownian motion so that less randomization in orientation occurs, thereby optimizing chemotaxis behavior. (van Duijn, Keijzer, and Franken 2006, 165)

To be sure, the kinds of explanation embraced by theorists of autonomy will differ significantly from the computational models of traditional symbolic cognitive science. But, all in all, they will need to relate to *some* kind of information processing. Otherwise, it is difficult to see how cognition should be possible; for, to repeat, processes that play no role in the transformation or communication of incoming information would hardly deserve to be called cognitive. An activity is cognitive insofar as it is sensitive to how the world is, and such sensitivity requires that there exist reliable processes capable of detecting information. It would be a mistake, however, to suppose that cognition consists simply of building maximally accurate representations of input information—though, admittedly, traditional information-processing theories of cognition focused predominantly on perceptual inputs. What these theories overlook is that the gaining of knowledge is a stepping-stone to achieving the more immediate goal of guiding behavior in response to the system's changing surroundings. The same mistake was made in the field of naturalized semantics, where "to represent" was analyzed either as "to stand in a natural relationship" or "to encode information." In both cases, as I argue in chapter 4, the notion of representation becomes deeply problematic. In particular, it is no longer possible to say what job the representation has in the cognitive system (Ramsey 2007).

The fact that cognition in real biological systems is not an end in itself has several important consequences. First, being constrained by the system's goals, cognitive representations cannot be mere copies of perceptual stimuli (nor can the former reduce to the latter). Second, information processing needs to be supplemented by such activities as exploration and modification of the environment. Third, information-processing mechanisms depend for their existence on a number of noncomputational structures. An explanation that ignores these structures would be incomplete.

That being said, one should still bear in mind that a general denial of the role of information processing in cognition, suggested by the dynamical approach, would lead to the conclusion that models in computational cognitive science cannot be genuinely explanatory; if they are predictive, it has to be a fluke. This, however, sounds like throwing the baby out with the bathwater; I will argue in this book that at least some computational theories are perfectly fine as explanations. Also, alternative and completely noninformational explanations of the very same phenomena do not seem to be forthcoming at all.

Many philosophers of science who set forth normative accounts of explanation do so without first analyzing how scientific inquiry is actually conducted. They often take various researchers' methodological pronouncements at face value and end up describing expressions of wishful thinking about science rather than science itself. This is especially tempting when the discipline in question—such as cognitive science—is relatively new. In fact, a considerable number of arguments against computational modeling in cognitive science rely heavily on general principles and thought experiments, as if no progress had been made since the publication of such programmatic books as *Plans and the Structure of Behavior* (Miller, Galanter, and Pribram 1967). True, when formulating the idea of accounting for behavior in terms of TOTE (test-operate-test-exit), Miller, Galanter, and Pribram did not have much empirical evidence to support it—but that has changed. Now, over forty years later, we see that cognitive science has collected a substantial body of empirical evidence and developed an impressive number of detailed models of various phenomena, so turning a blind eye to this wealth of new data and concentrating on "the first principles" is hardly excusable.

The truth is also that scientists are usually better at investigating the world than they are at speculating about the nature of their work. This is a second reason for not accepting their programmatic manifestos or introductory remarks on mere faith. Instead, we should look at how real science is done.

To this end, I will perform four brief case studies that show how computational explanations function within four diverse research traditions in cognitive science, namely (1) classical computational simulation, (2) connectionist modeling, (3) computational neuroscience, and (4) radical embodied robotics. The theories introduced in this chapter will recur throughout the book.

The examples have been chosen to represent classical approaches as well as the more recent research programs that stress the importance of neural (or, generally, implementational) data. I start with the work of Newell and Simon (1972), who explain cognition by positing high-level cognitive heuristics. Another study that falls into the traditional camp is David Marr's seminal book on vision (Marr 1982). Marr famously denies that the neuronal details are relevant to research on cognitive abilities (though, ironically, his own work draws on their intricate knowledge; note also that he is sometimes classified as a connectionist—see Boden 1988, chapter 3). (I examine Marr's claims in the context of levels of explanation in chapter 3.)

The second example is a study by David Rumelhart and James McClelland (1986), who model the acquisition of the past tense of English verbs. Although there is no shortage of other material, I focus on a fairly old paper from this research tradition, because classical studies tend to guide further research by showing students how proper work is done.

Computational neuroscience employs so many diverse methods that it is hard to find (nonconnectionist and nonclassical) modeling techniques applicable to numerous task domains. However, Chris Eliasmith's Neural Engineering Framework (NEF) is a recently developed strategy that may serve as a nice contrast to older programs expressed in the classical papers; this is especially true because it relies on certain dynamical properties and exhibits sensitivity toward problems raised by the proponents of dynamical models.

The choice of a paradigmatic example of radical embodied robotics (characteristic of the enactive research program in cognitive science) proved the most difficult. Given the preference of many roboticists for an engineering attitude—one that looks at successful solutions rather than accounts for the behavior of real biological systems—it would be unfair to argue that their robots fail to be explanatory. There are biological systems, however, that are actually being explained by robotic means nowadays—for instance by Barbara Webb. She also adheres to the principles of mechanistic explanation, which makes her work an especially interesting illustration for the chapters to come.

2   DONALD + GERALD = ROBERT

Allen Newell and Herbert A. Simon were among the founding fathers of cognitive science understood as a scientific discipline, and they are also considered to be important proponents of computationalism. Let us look at their research and see how computer programs are supposed to explain cognition.

In a study on cryptarithmetic problems, such as finding which digits correspond to letters in equations of the form SEND + MORE = MONEY or

DONALD + GERALD = ROBERT, Newell and Simon explain the subjects' performance by providing a description of an information-processing system (Newell and Simon 1972) formulated as a computer program capable of solving the problem in the same way as an individual subject. The proposed explanation focuses on individual performance and should be sufficiently detailed for an interpreter of the description to be able to perform the task (ibid., 11). In effect, what they offer is a *microtheory*—an account of a particular performance of a single agent (ibid., 165).

The performance is represented within a specific model of information processing known as an information-processing system (IPS). Its architecture embodies a number of psychological hypotheses about human problem solving, including limited short-term memory (ibid., 20–21) and the serial, or sequential, nature of high-level processing: "The problem solver may *see* many things at once; it only *does* one thing at a time about them" (ibid., 89). The IPS has receptors and effectors, a processor, and memory; it's clear that its structure is motivated by cognitive theory rather than theoretical computer models. (Only later did Newell [1980] demonstrate that the IPS is a universal machine.) To solve a task, the IPS transforms input symbol structures into output symbol structures that are accepted as a solution. Symbols are understood as entities that can designate other symbol structures (rather than referring to things external to the system, Newell and Simon 1972, 24).

Transformation of symbols is described in terms of production rules that may be expressed as conditionals whose antecedents give the criteria that have to be fulfilled, whereas the consequents give the actions to be performed (ibid., 32). A simple production might be, for example, "traffic-light red → stop" or "traffic-light green → move." A set of such productions, together with an entity capable of their interpretation and execution, is called a production system. This is why the IPS is a species of production system.

Before analyzing a subject's individual performance on the task, Newell and Simon hypothesize about the possible problem space, or spaces, in which a solution to the problem might be sought. As there are a number of ways in which problem spaces can be represented, the authors meticulously consider various options, such as basic, augmented, algebraic, and so forth. The empirical material is drawn from the verbal report of a subject who solved the task, and it's interpreted informally in order to reconstruct his or her problem space (ibid., 168–169) and build a graph that specifies the steps in the transformations:

> A problem behavior graph was constructed from his [the subject's] protocol; a set of operators was defined to account for the transformations of his state of information as he moved through the graph; a set of productions was proposed to account for his choices of moves or directions of exploration; and the details of fit were scrutinized between the production system and the PBG [problem graph behavior]. All of this analysis shows how a verbal thinking-aloud protocol can be used as the raw material for generating and testing a theory of problem solving behavior. A sizable fraction (75 to 80 percent) of the units of behavior was accounted for by the production system, and our detailed scrutiny of the discrepancies gives no reason for supposing that the remainder of the behavior is intrinsically incomprehensible or random. On the contrary, most of the inadequacies of the model appear to be due either to the lack of a detailed account of attention and memory mechanisms or to missing data. (Newell and Simon 1972, 227)

The end result is a production system whose performance matches the verbal protocol. The obvious problem with microtheories, it seems, is that they may not apply beyond their original scope. A naïve model constructed in terms of a production system might fail to account for the performance of other individuals on the same task and thereby defeat the very purpose of explanation. A model that simulates a single subject's behavior must be built in a way that allows for generalization. Surprisingly enough, Newell and Simon's production system fulfilled this requirement: the performance of other subjects turned out to be describable and predictable within the same system, and individual differences among the subjects could be captured as the system's parameters (ibid., 301).

There must be a trade-off between the depth of a simulation (or the degree to which it matches the verbal protocol) and the breadth or generality of the theory. Newell and Simon's explanatory method is no exception. Interestingly, however, testing showed that when the same IPS was evaluated on data extracted from eye movements, the match between the data and the system was much higher than when it was compared against verbal protocol alone (ibid., 326–327) despite the much finer grain of the eye-movement data. Because the system had turned out to be less than perfectly accurate when presented with verbal data, the researchers might have been tempted to exclude the productions that did not correspond with the verbal protocol. This would have actually crippled the model; the lower accuracy it displayed earlier was de facto a corollary of the fact that verbal reports usually contain gaps.

Verbal reports do not mirror thinking processes exactly. For various reasons, people sometimes failed to verbalize during the task, and some of the verbal protocol did not correspond to any steps of the task whatsoever. Nevertheless, the program—specified as production rules—is explanatory both of the original subject's behavior and of the behavior of other people on this task.

This case reveals several interesting features of an explanation in terms of traditional symbolic computation. The theory describes individual performance, which is conceived of as a search in a problem space, where the (possibly heuristic) steps are represented as transformations. A simple architecture of an information-processing system is postulated to account for symbol-transformation abilities that are supposed to explain human performance on cryptarithmetic tasks, and the level of detail given allows computer implementation.

However, the heuristic models do not take into account any realistic neurological constraints (the theory abstracts from them), nor is it claimed that the problem spaces of individual subjects correspond in any straightforward fashion to their lower-level neuronal realization. Only the cognitive agent's input/output behavior at the symbolic level is modeled (Simon 1993). Though certain internal mechanisms are posited (the architecture and operations of the IPS, productions, symbols, etc.), there is no way to test them *directly*. They are chosen solely on grounds of parsimony and the capacity to describe the input/output behavior in a way that is constrained by known psychological limitations, such as the capacity of the short-term memory (Miller 1956) or the number of high-level operations per second (Newell and Simon 1972, 808–809). The role of such psychological (or biological) evidence is to limit the space of possible models of cognition by excluding, say, the ones that posit capacity of the short-term memory that goes beyond seven plus-or-minus two meaningful chunks; it does not, however, allow for any kind of *direct* testing of computer simulations. In other words, the explanation shows how it is possible to solve certain tasks, yet it is still possible that the underlying architecture works quite differently in the actual system.

On the one hand, Simon and Newell are quite convinced that the regularities they observe at the symbolic level are robust and largely independent of the neural detail; on the other, they stress that they do not defend antireductionism or holism, and they advocate a view that "by virtue of the complexity of the phenomena, the reduction of stimulus-response regularities to neural processes will have to be carried out in several successive steps, not just a single step" (ibid., 875–876). In contrast to philosophical functionalists of the day, they do not argue for complete autonomy of psychology, but they do not want to embrace greedy physiological reductionism either. The most charitable reading of their defense of the symbolic-level posits would be that it is largely, but not completely, independent of the lower levels of organization; the stability of the higher level depends on there being few degrees of freedom at the lower level that behaves in a "rigid" way (Simon 1993, 639).

It is also worth noting that Newell and Simon do not analyze perception or motion using these methods, and they allow that sensorimotor processing might be parallel in nature (Newell and Simon 1972, 89). Exact timing of the task is not essential for its performance either, so it's left out of the scope of the theory. At least for these reasons, their theory may fall short of being a universal computational methodology in cognitive science. For my purposes, it is important to see both how the computational model is built and that it is basically a high-level simulation of behavior rather than a specification of the internal low-level structure of the cognitive agent.

3   A Network for Past-Tense Acquisition

The model of the acquisition of the past tense for English verbs developed by David Rumelhart and James McClelland (Rumelhart and McClelland 1986) is one of the most influential connectionist explanatory simulations. It achieves two goals: It shows that parallel distributed processing networks provide an alternative to the explicit but cognitively inaccessible rules postulated in a cognitivist explanation of past-tense acquisition, and it demonstrates how the training of a network might correspond to the development of mental skills in humans. The process of learning in a simple, two-layer neural network reflects three specific stages of past-tense acquisition in preschool children.

Let us start with the natural process to be modeled. At the beginning, a child who is learning to talk uses a relatively small number of past-tense verb forms—most of which are irregular. At this stage, he or she makes very few mistakes. The next phase is characterized by a rapid expansion of the child's repertoire of regular verbs and their past forms; at this point, he or she can create the past forms of unknown verbs, but at the same time the child overgeneralizes the newly acquired rules and occasionally produces incorrect versions of irregular verbs (e.g., using "comed" or "camed" as the past form of "come").[2] Hence, the learning curve is U-shaped, which makes the phenomenon of past-tense acquisition particularly interesting as a respite from the otherwise unrelenting tendency toward improvement on cognitive tasks that children exhibit as they age (Shultz 2003, 94). In the third and final stage of past-tense acquisition, which persists into adulthood, regular and irregular forms coexist.

The model is constructed so as to reflect these stages in the process of training the network; the results of the subsequent phases of training are supposed to resemble the way the typical child acquires the skill of changing the verbs into the past tense. In other words, the connectionist model, in contrast to Newell and Simon's cognitivist theory, does not correspond to a particular individual's performance: it is general from the very beginning, and it accounts for new verbs for this reason. Another important difference between the two approaches is that the connectionist model accounts for stages of acquisition rather than conscious steps of problem solving.

However, both models share certain important features. Even though Rumelhart and McClelland use a biologically inspired computational mechanism, their representation of the problem does not appeal to the way the human neural system works. Instead, they rely on Wickelgren's how-possibly analysis (Wickelgren 1969),[3] rendering each phoneme as a structurally rich pattern referred to as a *Wickelphone*. This solves an important problem that does not arise for symbolic theories but plagues connectionism (Bechtel and Abrahamsen 2002, 123)—namely, how to encode strings of information in a way that preserves order. The rub is that the input nodes of a connectionist network encode individual pieces of information by assuming on/off values corresponding to the presence or absence of a given feature—regardless of the feature's place in a sequence. Consequently, if a network simply detected unordered sets of phonemes, it would not distinguish between the sound profiles of, say, "came" and "make" (as they are made up of the same three phonemes: k, A, and m). Wickelphones remedy the situation because they are context-sensitive: each Wickelphone is an ordered triple consisting of the phoneme itself together with its predecessor and successor. For example, if we use "#" to indicate the word boundary, the verb *came* would be represented as $_\#k_A$, $_kA_m$, $_Am_\#$.

A major drawback of using this encoding method is that the number of Wickelphones is quite large. Supposing there are thirty-five individual phonemes in the English language, the number of Wickelphones must easily exceed 42,875 (i.e., 35^3, ignoring the Wickelphones containing the boundary sign). They are translated into an unordered set of properties the authors call *Wickelfeatures* (Rumelhart and McClelland 1986, 240). The translation reduces encoding redundancy, and the number of units in the network, to a manageable number (Bechtel and Abrahamsen 2002, 124), and it allows for a representation that provides a good basis for generalizations about the corresponding aspects of a verb's past and present tenses. Verbs, therefore, are not represented as strings of phonemes, but as an unordered set of these structural features encoded by the nodes of the network. To recover phonetic information, the authors use another neural network (see figure 1.1). Moreover, instead of feeding the network real linguistic data consisting of complete statements, the authors presented it

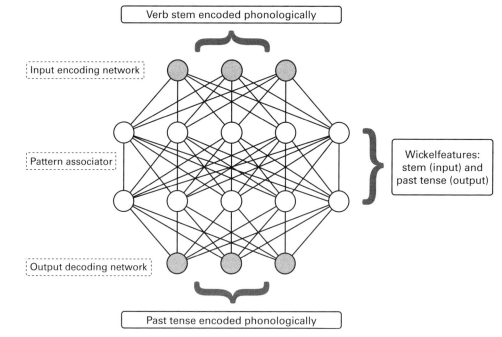

Figure 1.1
The structure of Rumelhart and McClelland's model.

with isolated phonological representations of verb forms (Rumelhart and McClelland 1986, 222). This is a significant simplification, for, unlike children, the model does not have to detect the verbs in a string of words. Children do not learn isolated verbs; they learn them as parts of sentences. More importantly, they do not learn unordered sets of features, and they do not seem to face the task of recovering strings from bags of features.

The authors also discuss the trade-off between the ability to detect various base forms of verbs that their Wickelfeature representation provides and the possible amount of generalization in the network (ibid., 238–239)—but in the human brain, this kind of trade-off does not exist. The trade-off is a psychologically irrelevant side effect of the implementation. The output of the model also exaggerates, to some extent, the differences between the stages of acquisition (ibid., 247).

The network was trained using the perceptron convergence rule, and the process had several stages. Initially, the network was presented with only ten high-frequency verbs in order to mimic the first stage in the development of past-tense morphology in children (on the assumption that a child cannot inflect a verb if it has not mastered its stem form—in fact, children hear more than ten verbs in stage one). In the subsequent phase, the network was exposed to 410 medium-frequency verbs, 80 percent of which were regular. While the authors justify this choice of training data by pointing out that it reflects the vocabulary explosion that occurs at some point in a child's development, it has been argued that the U-shaped learning curve for irregular verbs might actually be an artifact of the training procedure (Pinker and Prince 1988; Shultz 2003; see also Bechtel and Abrahamsen 2002, chapter 5.4.1, for more detailed data on empirically plausible estimations of verb frequencies during past-tense acquisition in children; I postpone critical discussion of this topic to chapter 5).

All these idealizations notwithstanding, the results achieved are impressive. The model's correctness is high (over 90 percent) not only for the known verbs, but also for previously unseen forms as well: 92 percent for regular verbs, and 84 percent for irregular ones (Rumelhart and McClelland 1986, 261). In this respect, the model is predictive of human behavior. Moreover, the simulation realized by the network could also generate correct hypotheses about the behavior of preschoolers, as reported in previous psychological research by Bybee and Slobin (ibid., 256; cf. Bybee and

Slobin 1982). The verbs that are difficult for the network to convert to past tense are also difficult for children. There is, in other words, an explicit focus on the predictive value of the model (some untested predictions are also mentioned, cf. Rumelhart and McClelland 1986, 272–273), and the authors stress that their explanation does not appeal to any explicit representation of linguistic rules. As they say, the model does not follow any rules perfectly, but neither do humans (ibid., 265). They summarize:

We have shown that our simple learning model shows, to a remarkable degree, the characteristics of young children learning the morphology of the past tense in English. We have shown how our model generates the so-called U-shaped learning curve for irregular verbs and that it exhibits a tendency to overgeneralize that is quite similar to the pattern exhibited by young children. (Rumelhart and McClelland 1986, 266)

Although it has been shown that what appears to be regular behavior might be explained without recourse to a set of explicit rules, McClelland and Rumelhart do not cite any neural evidence to support the hypothesis that this in fact is the case with past-tense acquisition in English speakers. In consequence, their model cannot be treated as a neurologically faithful account of brain mechanisms responsible for the performance of actual human subjects.

4   The Neural Engineering Framework

In modern computational neuroscience, various (quite often hybrid) computational models are used in explanations of neural phenomena. Some focus on the computational properties of single neurons (Izhikevich 2007); others concentrate on the level of the operations of whole groups of neurons (Rolls 2007). A range of methods and mathematical techniques are used to describe neural systems (cf. Dayan and Abbott 2001), and it is hard to talk about a single overarching methodology or theory (Eliasmith 2009, 346). There are classical symbolic models, like the ones proposed by David Marr (cf. Marr and Poggio 1976, where the explanation is regarded as complete once an algorithm that solves the task of accounting for stereo disparity has been sketched); there are also connectionist models. I will discuss the approach recommended by Chris Eliasmith, namely the Neural Engineering Framework (NEF), as it is sufficiently general for computational explanation of various neural functions.

The NEF is based on three basic principles (Eliasmith and Anderson 2003; Eliasmith 2009): (1) Neural *representations* are understood as combinations of nonlinear encoding and optimal linear decoding (this includes temporal and population representations), (2) *transformations* of neural representations are functions of variables represented by a population, and (3) neural *dynamics* are described with neural representations as control-theoretic state variables. Transformation, or simply computation, in the NEF is understood in the same terms as representation is—as nonlinear encoding and optimal linear decoding (Eliasmith 2009, 353). Representation is modeled as the nonlinear encoding of information in the temporal activity of neural networks. It might be considered as estimating the identity function, whereas computation "consists of estimating arbitrary linear or nonlinear functions of the encoded variable" (ibid., 354). In other words, when representing a variable, the system is concerned with decoding the value of that variable as it is encoded into neural spike trains. The decoders that compute any function of the encoded input, not just the identity function, are called *transformational decoders*. Eliasmith stresses:

This account of computation is successful largely because of the nonlinearities in the neural encoding of the available information. When decoding, we can either attempt to eliminate these nonlinearities by appropriately weighting the responses of the population (as with representational decoding) or emphasize the nonlinearities necessary to compute the function we need (as with transformational decoding). In either case, we can get a good estimate of the appropriate function, and we can improve that estimate by including more neurons in the population encoding the information. (Eliasmith 2009, 354)

The whole framework, though it features certain theoretical entities, such as decoders, which are not directly measurable, shows significant predictive success. According to Eliasmith, elements such as decoders are justifiable because of the critical role they play in determining connection weights in NEF models. Connection weights *are* measurable (though this is difficult). More importantly, these weights determine the tuning sensitivities of individual cells, the dynamic properties of their responses, and overall network behavior. Each of these is measurable and has been explained and predicted by NEF models across many neural systems (Kuo and Eliasmith 2005; Singh and Eliasmith 2006; Fischer, Peña, and Konishi 2007; Litt, Eliasmith, and Thagard 2008; Liu et al. 2011; MacNeil and Eliasmith 2011).

In some ways, the NEF is similar to traditional cognitive science: cognition is the processing (transformation) of representation. An important addition is the control theory, and Eliasmith emphasizes that it is crucial that representations play the role of control-theoretic state variables. In his opinion, only with control theory is realistic decomposition of the system into its explanatorily relevant component parts possible (Eliasmith 2010). The advantage of control theory is that it allows the inclusion of a real-time dimension in the dynamical description of the system; the state variables can be easily mapped onto the underlying physical systems given the other principles (in contrast to some dynamical equations).

One of the explanatory models based on the NEF concerns the path integration used by the rat navigation system (first presented in Conklin and Eliasmith 2005; for a less technical description, see Eliasmith 2009). Rats are known to return to their starting location after exploring, in a somewhat random fashion, the environment; they need not take the same route back (see figure 1.2). They are able to return to the starting position

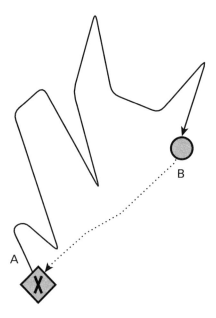

Figure 1.2
Rats are able to return directly to the starting point (A) after exploring the environment in the search for food (B). The return path is symbolized as a dotted line.

even when the only cues available are their own movements (so-called idiothetic cues); the widely accepted hypothesis is that the rat represents the environment on a mental (or neural) map and updates its location on that map as it moves about. The rat is therefore one of the animals capable of dead reckoning—or path integration, as this navigational capacity is called in biology. (It was Charles Darwin who hypothesized in 1873 that animals have inertial navigational systems; for a modern summary, see Shettleworth 2010, chapter 8.) Modern neuroscience vindicates Edward Tolman's idea of cognitive maps posited in rats (Tolman 1948)—especially since O'Keefe and Nadel (1978) located the maps in the hippocampus. There are two map systems in the brain: one based on place cells in the hippocampus, and the other on grid cells, border cells, and head-direction cells in the parahippocampal cortex (Derdikman and Moser 2010). As the behavioral and neuroscientific evidence is rich, models of navigation, including those of path integration in rats, abound in computational neuroscience (for reviews, see Trullier et al. 1997; Redish 1999; McNaughton et al. 2006).

Neurophysiological evidence shows that the representation can be thought of as a bump of neural activity centered on the rat's current estimate of its location. To computationally explain the behavior on a neural level, Conklin and Eliasmith take the known physiological and anatomical properties of the neurons in the areas active during the rat's navigation task and show how their organization could implement a mechanism that reproduces the observed behavior. They start out by representing the bump of activity as a two-dimensional bell-shaped function. The nonlinear encoding is then posited (the details are not important here) based on both the neural model used by the authors and the response of neurons observed in the active areas.

The next step is a sketch of a high-level mechanism responsible for navigation to the starting location based only on self-motion velocity commands from the vestibular system (the mechanism is a dynamical/control system). There is a stable function attractor that holds the bump of activity in the same location in the case of no movement or moves the bump in the case of self-movement; the representation encodes the position in the environment, and it is updated based on velocity commands from the motor system (only in two dimensions). This transformation of the neural representation is biologically realistic:

In short, we calculate the feedforward and feedback connections between the relevant populations of neurons by incorporating the appropriate dynamics, decoding and encoding matrices into the neural weights. Interestingly, we find that the resulting weight matrices have a center-surround organization, consistent with observed connectivity patterns in these parts of cortex. (Eliasmith 2009, 358)

The representation is thereby embedded into the postulated control system according to principle (3) above: it is a state variable of the system.

The model was applied to generate numerous simulations resulting from the activity of four thousand neurons. It used a novel, biologically plausible mechanism of a spiking attractor network composed of leaky integrate-and-fire neurons (for more background, see Eliasmith 2005a). When completing a circular path, there was a slight drift error (11 percent), which was much smaller than in previous models using three hundred thousand neurons (Samsonovich and McNaughton [1997], who had an error rate of 100 percent). It is possible to replicate available data (especially concerning phasic phenomena in rats) and make three predictions: (1) Cells will be velocity- (and position-) sensitive, (2) cells in the navigation system under explanation will have the same location in different environments, and (3) the head direction of the rat will not affect its ability to return to the starting location. The third prediction contradicts earlier models of this phenomenon. Overall, the predictions of the model are both behavioral (relating head direction to integration bias) and neural (predicting specific relations between neural tuning curves and the rat's environment).

This model, in contrast to classical cognitivist and connectionist simulations, is based on empirical neurophysiological data and stresses the significance of lower-level detail. Yet it can recreate high-level behavior as well. Since the crucial mechanism, the dynamical/control system, treats the transformed, or computed, representation as a variable, the whole explanation is also computational. The last claim might seem contentious, as Eliasmith (2010) seems to contrast the computational modeling that cannot yield, in his opinion, realistic decompositions of the system, with the control-theoretic modeling; for this reason, computational modeling is thus rejected. However, his critique focuses only on traditional symbolic computational models—the ones that are understood as completely independent of physical implementation. Such models are theoretically grounded only in metaphor, which is problematic (Eliasmith 2003). In subsequent chapters, I will argue that symbolic computational models that are completely divorced from their implementations are indeed less explanatory, but that they are not the only kinds of computational models in cognitive science. And they are surely not metaphorical.

The NEF seems to bridge the gap between the low-level neuronal activity and high-level cognitive mechanisms. Interestingly, the NEF can also be used to replicate the famous results of the Wason task in cognitive psychology (Eliasmith 2005b; cf. Wason 1966). At the same time, the NEF is "unique among approaches to the behavioral sciences in that the elements of the theory are (very) directly relatable to observables, and hence the models stemming from the NEF are highly testable" (Eliasmith 2009, 362).

5   Embodied Robotics

One of the uses of robotics is to explain the behavior of animals. As the complexity of many animals surpasses the complexity of today's robots, the stress is on simple creatures such as insects. It is currently possible to build detailed robotic models that will explain insect behaviors in the physical environment rather than just simulate the insect's neural network on a computer. There are currently two strands of research in robotics relevant to cognitive science. On the one hand, there is research on *animats*, or possible creatures, which is supposed to provide insight into the principles of cognition (and has roots as deep as Tolman's [1939] "schematic sowbug"); on the other, there is robotic simulation of animals intended as explanations of real biological systems. While the claim that animats are genuinely explanatory of any biological systems is controversial because animat studies rely on models that do not describe, explain, or predict the behavior of any existing entity (for further criticism, see Webb 2009), the latter approach appears to be a viable strategy for cognitive science (Webb 2000, 2001). To be sure, real organisms are much easier to test than conceptual creatures, which is why I focus on them—although, admittedly, animats are much more popular (among philosophers as well as cognitive scientists). I do not mean to suggest that purely synthetic methodology is useless; thought experiments and theoretical modeling may be systematic and well-motivated.

Although it is possible to model some features of higher animals on robots, simpler insects may be simulated in much more detail. One of the insects investigated in this way is the cricket (Webb 1995; Lund, Webb, and Hallam 1997; see Webb 2008 for a review). Crickets display several interesting behaviors; one such behavior is phonotaxis, or the female's ability to walk toward the location of the male based on the male's calling song. The first step in simulating phonotaxis is to investigate the cricket's behavior. Empirical data, however, are rarely sufficient to build a complete robotic mechanism (Webb 2008, 11). The carrier frequency of the male cricket's song, which is used by the female to detect conspecifics, is around 4–5 kHz; the ears of crickets, located on their frontal legs, consist of a pair of tympani and vibration receptors. Crickets have a peculiar auditory system in which the two ears are connected by a tracheal tube to form a pressure difference receiver (as it is called in engineering). This gives them good directionality but only for the specific range of frequencies of the calling song they need to localize (see figure 1.3). The same solution was mimicked on a robot (for details, see Webb 2008, 10). The sound information for other frequencies does not have to be filtered out by the cricket, as the very physical setup makes the calling song frequency easier to localize—other frequencies are still audible but harder to locate. Information processing in this case is not really essential to the explanation of how filtering is done.

Mate-finding behavior is not simply a matter of approaching the loudest sound—females select the sound that has the right temporal pattern. Early studies suggested that filtering is needed to recognize the pattern, and the sound's direction is used to determine the direction of the turn of the legs of the cricket. This theory was later called into question (Webb 2008, 29).

Barbara Webb and her collaborators hypothesized that the significant cue for the filtering could be the sound's onset. This idea was embedded in a simple neural circuit connecting the auditory neurons with motor neurons—the left auditory neuron excited the left motor neuron while inhibiting the right motor neuron, and vice versa. As a control mechanism, the circuit was able to reproduce a number of behavioral experiments on crickets (Webb and Scutt 2000). Although the posited mechanism can predict the turning behavior, it is not explanatory of the internals of the neural system of the cricket. In fact, the known physical characteristics of the cricket neural system contradict the simple circuit model: "The use of both excitatory and inhibitory synapses from a single neuron is already not consistent with neurobiology" (Webb 2008, 28). Instead of accounting for all the known details of the neural system, Webb suggests that, given

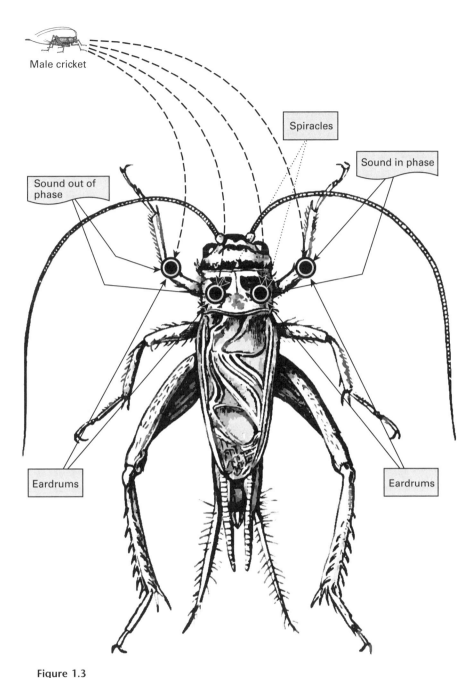

Male cricket

Spiracles

Sound in phase

Sound out of phase

Eardrums

Eardrums

Figure 1.3
Female crickets localize the males with their specially structured ears. The sound is out of phase on the opposite side of the eardrum, which means the eardrum has a larger amplitude of vibration on the side closer to the male. Top cricket: Source: WikiMedia Commons: http://commons.wikimedia.org/wiki/File:Cricket_%28PSF% 29.png. Bottom cricket: Source: Adam Sedgwick, David Sharp, and F. G. Sinclair, *The Cambridge Natural History*, vol. 5: *Peripatus, Myriapods, Insects*. Edited by S. F. Harmer and A. E. Shipley. London: Macmillan, 1895, p. 330.

its explanatory goals, the model should include only those elements that are assigned a role in the mechanism producing the behavior under investigation. Her subsequent experiments are more realistic toward the neural data that have been collected since the early attempts to model cricket phonotaxis (see Webb 2008).

Robotic models are supposed to supply mechanistic explanations of behavior. Task analysis is only one of the components of a complete explanation. Researchers also focus on features of the environment and on the physical makeup of the cognitive agents. In the case of insects, cognitive processing cannot be the sole explanatory factor; the mechanisms also include certain properties of the sensors, actuators, and the physical environment. Physical interaction, rather than just simulation, allows us to discover certain features that simplify the cognitive task for the organism in question. Nevertheless, the strategy is not constrained to the lower level of organization; it can also account for higher-level tasks (such as mate-finding). The data to be explained come from experiments, and predictions yielded by robotic models can be tested on real animals. What is interesting in this connection is the fact that computational processes posited by a computational explanation are among a number of causal processes that contribute to the overall behavior of an animal. Rather than deny the role of neural computation, behavioral roboticists use it to explain animal behavior in the context of a bigger, mechanistic explanatory project.

The model under consideration concerns a fairly simple phenomenon, and it might be objected that it fails to be complex enough to qualify as "cognitive" (Adams 2010). It is already included in introductions to cognitive science, so various authors do find it sufficiently interesting to analyze as an important example of embodied cognition research (Clark 2001a, 103–108; Bermúdez 2010, 437–438). Yet as the very claim that embodiment is essential for cognition may be contested (and Adams, as an opponent of the extended-mind thesis, may object that it would be begging the question on my part), we need some independent reason for thinking that the phenomenon is cognitive enough to be juxtaposed with other models here. I think the most important difference between the robotic cricket and others is that there is no use for genuine cognitive representations here, and the robot is merely reactive. For those who think that the mark of the cognitive is representation (e.g., Rowlands 2009, Adams 2010), it might mean that there is no reason to call information processing in the cricket "cognition." But it is surely cognitive in the light of the "minimal cognition" program, where bacteria and plants are analyzed as cognizers. There is interesting online adaptive behavior in these cases to be explained, not exclusively but importantly, with information-processing mechanisms.

6   Too Early for a Funeral

As the preceding glimpse at computational explanations shows, the reports of computationalism's death have been vastly exaggerated (protestations to the contrary notwithstanding). Computation is posited in order to explain and predict a range of cognitive phenomena. The scope of the explanations varies with the strategy. For the classical cognitivist and (to some extent) connectionist models, the empirical data concern only the cognitive task. The gory details of the neural systems are left aside. In the models that start with the lower-level processes, the internal processes can also be explained, at least to some extent, computationally. It should be noted, however, that embodied robotics and computational neuroscience do *not* preclude task analysis; the modeler has to know what kind of higher-level behavior is supposed to emerge in the first place. In fact, most explanatory strategies do analyze the behavior of a system in terms of input and output data (where the output might be expressed not only in terms of "decoding" the input but also as motor activity). The examples also show some focus on how the input data are represented and how efficient the process is that deals with the data. For instance, complex symbolic processing hardly presents a plausible hypothesis about cricket mate-finding behavior.

Obviously, it could be argued that my four examples are biased. There is, admittedly, some prejudice in favor of work that is empirical rather than just programmatic. In this chapter, I did not report any criticisms of these models; this is because, in order to evaluate the significance of criticism, I need to develop and justify a theory of computational explanation in cognitive science (which is the focus of chapter 3). Fear not—critical evaluations, such as the influential polemic by Pinker and Prince (1988), will come to the fore later (in chapter 5). There are also explanatory strategies that I did not cover here. More specifically, I postpone the discussion of dynamical systems until chapter 5, for it requires a more explicit analysis of the concept of computation. I do think, however, that most mainstream modeling work is similar to the examples I have briefly discussed.

What is immediately obvious is that, as far as explanation is concerned, the notion of the universal Turing machine is irrelevant (see also Sloman 1996). Although some researchers argue that formal analysis is critical to finding the most efficient algorithms (for a strong supporter of such a position, see Marr 1982), the process of assessing efficacy boils down to estimating the computational complexity of the algorithm (see van Rooij 2008 on the importance of the condition of tractability for computational theories of cognition). This does not require building a complete description in terms of the Turing machine. Contrary to what is sometimes suggested by philosophers, cognitive scientists employ a wide range of computational models, and they usually do not refer to the Turing machine at all.

Empirical researchers posit information processing to explain the behavior of cognitive systems, and that seems to be the main driving force behind computational explanation.

We talk of information processing whenever some processes rely crucially on information. In most cases, this happens in complex systems in which information is an essential factor in the overall behavior. For example, it sounds a bit bizarre to say that a thermometer processes information about the temperature if we refer to a mundane analog thermometer with a glass tube containing some colored ethanol. Yet a thermometer with a digital display, and especially one driving the behavior of a thermostat, seems to be processing information in the paradigmatic way. The analog thermometer is a system that merely reacts to the environment, creating natural signs (or signals), whereas the digital thermometer encodes these natural signs and uses them to accomplish something. It seems that the only way to account generally for such information processing is to appeal to the notion of computation.

Information processing need not be digital in nature; a thermometer might drive the thermostatic device in an analog way (e.g., some metal parts, such as internal thermometers, can be sensitive to temperature changes and bend accordingly). In computer science we also talk of analog computation. I propose that the notions of computation and information processing be used interchangeably. This will allow us to translate all talk of information processing into clear computational terms.

2 Computational Processes

In this chapter, I analyze what it is for a physical process to realize or implement a computation. I argue that the computational theory of mind is to be understood literally. Its literal rendering requires that we know exactly what it is for a physical entity to implement a computation. I review the philosophical theories of implementation that have been offered so far and show that they cannot handle a number of important difficulties. As an alternative, I present a mechanistic account of implementation of information-processing systems—one that cannot be refuted by skeptical objections.

The purpose of the analysis I offer is to vindicate the view that computationalism is a viable theory with clear ontological commitments that cannot be trivialized by mere armchair arguments. The four cases of computational explanation sketched in the previous chapter posit theoretical entities, and the goal of a mechanistic account of implementation is to say what these entities are. The next chapter, in turn, shows how to test these theories and what it means to say that they are explanatory.

1 The Computer Metaphor, Computation, and Computationalism

In philosophy and cognitive science, the computational theory of mind is often called "the computer metaphor." This is unfortunate because it is far from clear what is meant by "metaphor" here. First, it might be taken to suggest that an electronic computer is understood as essentially the same kind of device as the brain; this is obviously implausible (Hunt 1989, 604). Second, one might read it as asserting a weak analogy between cognition and digital computation, and weak analogies are less explanatory than literal theories (Eliasmith 2003). Third, it is sometimes used disparagingly to suggest that people have always talked about the mind in terms of various technological metaphors, such as pumps, clocks, and steam engines, and the computer, which happens to be the most recent addition to this list, will eventually be replaced by something else (Daugman 1990). Being ambiguous, the conception labeled "the computer metaphor" becomes a moving target; it is difficult, if not impossible, to disprove or even undermine a metaphor because its advocate can always protest to have meant something else than the critic implies. It is much easier to establish the inadequacy of a literal description.

I will, therefore, take the computational theory of mind literally rather than figuratively. Like Newell and Simon (1972, 5) and Zenon Pylyshyn (1984, xiv–xvi), I believe that the literal treatment of computer models of the mind commits researchers to rigorous testing. In other words, computer models, rather than serving as mere demonstrations of feasibility or loose metaphors of cognitive processes, have to live up to the standards of scientific investigation when treated literally. Moreover, I believe there is a straightforward reading of the computational theory of mind that makes it nonvacuous, empirically testable, and methodologically strict. The claim is not that the mind is a computer, but that one of the essential levels of the cognitive system's organization is best described as information processing. It is primarily at this level, the level of mental mechanisms (Bechtel 2008a), that cognitive processes ought to be explained.

The current use of the term "computation," and of several related concepts (such as algorithm, program, digital, and analog), is not uniform; this is true despite a certain amount of overlap in the term's meanings. In short, what one psychologist, philosopher, or cognitive scientist means by "computation" is likely to be at odds with how another researcher understands it. One source of these discrepancies is that once the notion had been applied in a narrow sense, such as in the classical computational theory of mind (associated with symbolic models and the language of thought hypothesis[1]), many authors in a given field take it for granted that no other use is legitimate. Now, given that the term is featured within a number of conflicting theoretical frameworks from various research traditions, each placing its own constraints on the concepts it employed, terminological disagreement was bound to arise. The upshot is that there are many construals of computation on the market, and they do not necessarily depend on the same theoretical assumptions. My initial goal, then, is to dispel some of the conceptual confusion here.

To defend the computational theory of mind, theorists usually develop an account of computation that is geared toward theories of cognition (e.g., Cantwell Smith 2002). I think this is a mistake. First, there might be computational processes that have been overlooked by cognitive science; why should we decide a priori that cognitive science cannot make use of developments in computer science? Second, accounts of computation as presupposed in certain theories of cognition have already contributed to conceptual confusion. For this reason, if a psychological paper contains a denouncement of computationalism without any further specification, then it is not clear what is actually being rejected: is it any computational theory of cognition, or a specific one that presupposes, for example, that the medium of computation is one of quasi-linguistic symbols?

For the sake of clarity, it is best not to conflate the theory of implementation with the theory of computation as geared toward cognition (for a similar argument, see Fresco 2008). This is why I will first develop the broadest possible account of implementation—one that is intended to cover any computational process in the physical world and not just cognitive processes. Moreover, cognitive science is just one of many disciplines that explains computational processes. For example, it could turn out that there are some computational processes that do not occur in cognitive systems, but are nonetheless interesting enough to have a theory of implementation for them. Hyper-Turing computations occurring in Malament-Hogarth spacetime or in orbit around a rotating black hole are a good example: even though, for whatever reasons, there could be no cognitive systems around a black hole, we might still be interested in understanding what it takes for computation to be implemented in such space-time locations. Also, explaining the operation of computational artifacts does not require appealing to cognitive science at all.

The specific needs of cognitive science, which arise mostly because of its explanatory and predictive interests, will be addressed in chapter 3. This way, ontological issues are separated from epistemological ones that arise only for cognitive science (and related disciplines). Note that there is no similar confusion concerning the notions of explanation or prediction in science, so I do not need to develop a new general account of them in order to apply it to cognitive science. This is why the following chapter adopts a different strategy.

Even if computer scientists discuss the limits of computation, they usually do not aim to model physical processes and focus instead on computational solutions to various theoretical or engineering problems. They tend to use "computation" either in the intuitive sense or in the technical sense. Some take the latter to be equivalent with the now-standard notion of an effective algorithm, and others consider it to be more general and to cover other computational processes, such as analog computation on genuine real numbers. (One could even argue that there is considerable disunity in how people use the notion of computation in computer science as well, but I do not think the issue is remotely as severely confused as it is in cognitive science. For my argument, however, this does not matter.) This is why we need an additional theory of what it is for a physical process to be computational. For my purposes, the only thing that is important here is that the philosophical account of implementation should be compatible with any notion of mathematical computation used in computer science, mathematics, or logic. But it will not be deducible from mathematical theories of computation at all.

Computational processes are not just artifacts of human technology. Rather, like many other kinds of physical processes that fall under true quantitative descriptions, they constitute natural kinds for science. To be sure, there are other quantitative descriptions of these processes, be it in terms of statistics, control theory, or differential equations, but accepting one of them does not commit us to rejecting the rest. Provided that such descriptions are genuinely explanatory, the entities they posit cannot be dismissed as epiphenomenal or existing merely in the eye of the beholder.

I begin by presenting, with its strengths and weaknesses, an intuitive simple-mapping theory of implementation. I then show that no account of implementation that is based merely on mapping can escape devastating objections. Next I turn to a representational (or semantic) view of computation—by far the most popular among philosophers—that, contrary to appearances, cannot overcome the problems of the simple-mapping theory. However, there is a grain of truth in the semantic view, which can be saved by appealing to the notion of information. This notion leads naturally to the idea of information processing, which will be further constrained to develop an alternative, structural account of physical computation. The end result is an adequate mechanistic account. Readers interested in the positive theory rather than the discussion of weaknesses of traditional philosophical proposals may skip the fairly technical sections 2 and 3, and go directly to section 4.

2   Isn't This Just a Mapping?

Computational processes in the physical world have two salient properties: (1) their structure can be described in terms of a computation, and (2) they are physical. Hence, the intuitive idea that many philosophers and physicists endorse: a physical process is computational if, and only if, there is a computation such that the physical states the process consists of correspond to states of the computation in a one-to-one fashion (see figure 2.1). After all, a true computational description of a process must correspond to reality. This correspondence is usually framed in terms of isomorphism: the structures involved have to stand in identity relation to each other.

Simple and elegant as this idea appears, a moment of reflection suffices to discover its implausibility. For one thing, given that computers have plenty of physical properties that are not implicated in computation, it is hardly credible that physical systems have as few states as their corresponding abstract computational structures. Some physical states must be inessential vis-à-vis the system's computational characteristics; surely, the fact that, say, a sticker on my laptop computer is coming unstuck does not influence the performance of my word processor. So at least some physical states will have no computational counterparts (see figure 2.2).

The second complication is even worse. Given that physical states are individuated within a particular theory and that different theories can carve the world at different joints, one can artificially generate any number of physical states in order to map them onto one's favorite computation.

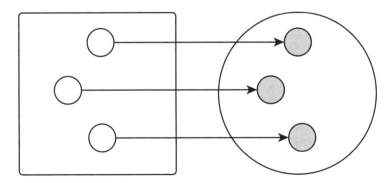

Figure 2.1
The physical states (on the left) are said to correspond to computational states.

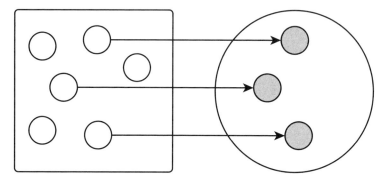

Figure 2.2
Some physical states (on the left) will have no computational counterparts.

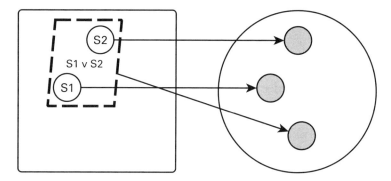

Figure 2.3
By logically combining the states of a physical system (on the left) that has fewer atomic states than a model of computation, a strict correspondence between the system and the model can be established.

So even if the physical system in question has *fewer* states than its allegedly corresponding mathematical construct, one can define as many physical states as one wishes by means of set-theoretic operators, such as union, or their counterparts in a logical calculus (see figure 2.3).

Intuitively, sufficient complexity of the physical system and a bit of mathematical ingenuity would be enough to show that the system in question implements any computation under the simple-mapping account. Indeed, Putnam constructed a proof that any open physical system implements an inputless finite-state machine (Putnam 1991, 121–125), which led him to the conclusion that "every ordinary open system is a realization of every abstract finite automaton." The purpose of his argument was to demonstrate that functionalism, were it true, would imply behaviorism (ibid., 124–125); internal structure is completely irrelevant to deciding what function is actually realized. Therefore, functionalism, if it is not to be conflated with behaviorism, ought not to endorse the view that the relation of realization is best conceived as a simple mapping of two structures. As Godfrey-Smith (2008) argues, there is a tension in classical functionalism: On the one hand, it argues for antireductionism in order to eschew the reduction of computational states to physical ones; but on the other hand, without reducibility, classical functionalism is always open to trivialization arguments showing that higher-level properties are just in the eye of the observer.

Putnam's proof is based on the idea that an ordinary, open physical system consists of many physical states. According to any external real-time clock, these states obtain for some time. So, if a computational model (Putnam uses an inputless finite-state machine, but the idea is generalizable) requires two distinct states that occur in some predefined sequence, we need to define these states as disjunctions: the first state will be a disjunction of some physical states, say s_1, s_3, and s_5, and the second will be a disjunction of all the other states (with no overlapping; see figure 2.4). It suffices that we get the required sequence of computation states from a set of physically individuated states by combining their descriptions with a logical operator of disjunction. The truth of the description will be always

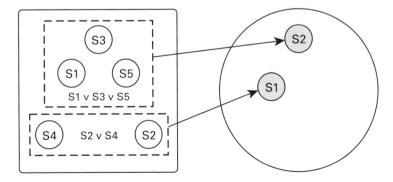

Figure 2.4
Logical combination of states gives rise to arbitrary entities (denoted with dotted lines) that reflect two states of the finite-state automaton.

preserved (after all, addition of disjunction is one of the standard rules of propositional calculus), so we are free to combine true statements any way we like—including the way that would make the evolution of the system clearly correspond to the formal model of the computation. Or so says Putnam.

The problem is that we would need to generalize this principle to any measurable feature of the world, thereby making any complex mathematical description of it a matter of arbitrary decision. Also, any physical system, be it a tornado, planetary system, or pail of water, could justifiably be called a computer. It's not just that there is one computation that may be ascribed to a physical system, which would not be as detrimental for explanatory purposes as making implementation a matter of completely arbitrary decisions. If there is a trivial computation that a rock implements, then it is not a tragedy for the theory of implementation (Chalmers 1996a). But if the rock implements all possible computations at the same time, then the notion of computation is utterly trivialized; there cannot be any explanatory or predictive gain in using computational models in cognitive science, or anywhere else for that matter, including computer engineering. Since the simple-mapping account as it stands lacks the resources to deal with this objection, its advocates have to come up with some additional constraints to prevent disaster. Shortly, although it may be mildly counterintuitive to many that everything implements some computation, computationalism would be truly bankrupt if one could justifiably say that any computation is implemented on just anything. But the simple-mapping account allows that any computation is implemented by anything.

To prevent such disaster, Copeland (1996), for example, suggests that honest mappings should be distinguished from nonstandard ones. He observes that nonstandard models of computational structures (1) have either been constructed ex post facto, or (2) introduce temporal specificity, as in Putnam's proof, or (3) appeal to material implications. Accordingly, he stipulates that honest mappings must support counterfactuals and not be devised ex post facto. (Putnam relied on the clock, but this is not essential; one could easily define additional physical states by using some spatial relations, which is why Copeland does not require that honest ascriptions not exploit temporal structure.)

The first weakness of Copeland's theory is the vagueness of the condition that the mapping not be constructed ex post facto. What should one

make of it? Obviously, to show that a finite state automaton F is instanti-
ated in a system S, Putnam needs to know some structure of S, and then
reassign the states of F to states of S by carving S at artificial joints. Why
would that violate the proviso against introducing mappings ex post facto?
Simply, because Putnam would have to know both structures beforehand,
including the real structure of S, and then reassign the states of S artificially.
The rub is that it is quite difficult to pin this requirement down in a general
way. Should that mean that an honest correspondence relation cannot be
known prior to empirical findings regarding the structure of S? This would
be too restrictive. For example, a person trying to reverse engineer an
unknown device cannot know its structure both in advance *and* posterior
to empirical research; unable to come up at once with a complete mapping,
he or she will typically have to take the device apart to see its components
and try different hypotheses. But as the general theory of systems clearly
shows, one cannot infer the complete structure of all complex systems
from their external behavior—even when one has complete knowledge of
the input and output of a black box and its atomic component parts
(Zeigler 1974)—so the complete structure has to be postulated rather than
discovered. Why would such postulates, if they were put forward as
hypotheses *after* dismantling the system into component parts, be less
honest than the ones that were offered in advance? Copeland might have
in mind that the mapping should be established not by freely reassigning
the spatiotemporal states of S to states of F when we already know how S
functions, but instead by accounting for the structure that we have already
discovered in S. This, however, is not really related to the formulation of
hypotheses beforehand or ex post facto, and we might *have to* assign some
structure ex post facto because of the lack of appropriate evidence; a theoretical postulate that the structure S is actually an implementation of F
might be fruitful then. What is wrong with Putnamesque ascriptions is not
that they are proposed ex post facto but that they ignore what we already
know about the system. But then, to make Copeland's constraint useful,
one needs to specify what kind of empirical knowledge is required to make
honest ascriptions. Yet this has clearly slipped Copeland's attention. As it
stands, his first criterion is either too restrictive or lacks content because
of its vagueness.

The requirement involving counterfactuals is not without its difficulties
either. It too could be satisfied, given enough ingenuity, by constructing a

function that takes into account all possible evolutions of the physical system. Such a mapping need not be sensitive to the system's internal organization; therefore, by stipulation, I could simply define the states of a physical system so that they always correspond to the computational model. The definition will be a bit more cumbersome, but the procedure of creating one is surely effective. Let me elaborate. A finite-state machine F will always have n states, where n is a finite integer. In the standard Putnamesque case, what we needed to secure is a one-to-one mapping of n states into x states of the physical structure. In the more sophisticated version, we will need to secure a one-to-one mapping between all possible evolutions of these n states, which is again a finite, though possibly a large number m, and all evolutions of the physical system. As we will not really look at the internal system organization anyway, we can simplify the task by starting from the particular number of states that the evolution of the finite-state machine comprises, which we will call k. After all, we can easily enumerate all the evolutions of F, as they are deducible from its definition. So the task reduces to always finding k states in a physical system (in a more difficult version, they would have to be in appropriate order). Instead of using the external clock, we can simply define the states by saying that in this particular temporal evolution, it comprised k parts that correspond to the states of computation. This is counterfactually true: for any input state of F, the appropriate evolution of F will correspond by definition to the evolution of S. Now, this might be still too easy for some, as there would not be systematic differences between various "runs" of F "implemented" by S; the system might as well stay physically the same. We could get around this as well by creating m slightly different assignments of evolutions of F to S. Other simple examples that satisfy Copeland's counterfactuality criterion but rely on arbitrary groupings of physical states are given by Scheutz (2001), who also analyzes them in a more formal fashion than I did above.

In fact, counterfactuality is almost a vacuous constraint if we do not specify some specific kind of structures of physical and formal systems that need to be appropriately accounted for in counterfactual assignments. Plus, some counterfactual statements about implementation are, at least for the standard electronic computers that I know of, false: computers break in nonstandard conditions, so the assignments should be true counterfactually only in a limited number of cases. Otherwise, my laptop is not a

computer because it stops working if immersed in water. Putnam's counterexamples can be patched to satisfy these criteria; the heart of the
problem is that the physical states are not the ones that support the causal
structure of the computational process.

This is not to say that it is logically impossible to develop a theory of
computation that would specify its requirements in terms of what may be
assigned to what, or what may be mapped to what. In other words, I do
not deny that it is possible to talk in a formal mode about implementation;
instead of referring directly to physical processes, one can talk of descriptions of these processes. But it escapes me in what way that would be an
improvement over simple talk of physical processes and their causal structure. It seems that the best available account of implementation in terms
of assignments was spelled out in measurement-theoretic terms (Dresner
2010). Dresner holds that implementation is basically a representational
relation—that is, that the physical can be described in terms of the computational as if the physical was measured in terms of the computational.
As a result, there are two basic requirements of such an account: (1) The
physical can be represented in terms of the computational, which is satisfied if a representation theorem is proved, and (2) a uniqueness theorem
must be proved showing that all mappings from the physical onto the
computational are simply notational variants (such as different scales on
a thermometer). In particular, all that is required is that there is a homomorphic (structure-preserving) mapping from the physical into the computation. Now, to say what is structure-preservation, Dresner would have
to go beyond representational theory of measurement, as this theory is not
interested in saying what *structure* is of interest for computational implementation at all. The problem is that his account still does not say what
is to be preserved, and although Searle's arbitrary mappings (see section 3
below) might be blocked, arbitrary Putnamesque disjunctions are still possible. Saying that there is homomorphism is simply not enough (Scheutz
[2001] presents several simple homomorphic mappings that do not constitute any computation in the physical system under consideration); when
we say what should be preserved in this mapping, we go beyond purely
formal characterization.

The simple-mapping account, though popular, including among physicists (e.g., Deutsch [1985] seems to assume it), is not the preferred way of
thinking about computation among philosophers. Isn't computation the

manipulation of symbols? Isn't it operating on representations? Jerry Fodor
suggests that the difference between computers and planetary systems is
that the latter do not represent the rules that govern their operation (Fodor
1975, 74). Let us investigate then the semantic and syntactic accounts of
computation.

3   Semantics (or Was It Syntax?) to the Rescue!

The formal symbol manipulation (FSM) account (Fodor 1975; Pylyshyn
1984), which many philosophers of cognitive science embrace as self-
evident, says, roughly, that computing is manipulation of formal symbols.
There are two notions that need to be unpacked here. First, it is not at all
clear what formality of symbols consists of (Cantwell Smith 2002); second,
the concept of symbol itself is also far from obvious.

How should we understand "formality"? Let me start from an intuitive,
nonproblematic reading—that is, that formal symbols need *not* refer to
anything. But there are other semantic properties one might think that
formal symbols do not possess. The problem is that it is hard to enumerate
them all without committing oneself to a particular theory of meaning.
The most noncommittal reading would therefore be that formal symbols
are symbols that do not need be *nonformal*, which is a slightly convoluted
way of saying that their formality is sufficient for computation but not
necessary. From the semiotic point of view, a formal symbol is a sign whose
function is determined merely by its form (which is not to be identified
with its shape, as not all vehicles have visual forms; a form is an abstract
property of a vehicle).

Note that if we define formal symbols as devoid of nonformal properties
once and for all, then, according to FSM, the words that I type right now
on my word processor would not refer either, which is absurd (hopefully!).
They just need not have any nonformal properties (I can type blahgrahad,
which does not mean anything, at least in any of the languages that I
speak). There are other ways of understanding formal symbols, such as, for
example, "independent from semantics" (see, e.g., Cantwell Smith 2002).
But such construals require explication of what we mean by "semantics,"
and this could make the FSM account dependent on semantic notions,
which is what FSM explicitly tries to avoid. So it is best to keep the definition of formal symbols maximally noncommittal in this regard.

Another problem with the formality-of-symbols claim is that it lacks proper quantification. Does it assert that *all* computation involves *only* formal symbols, or that *all* computation involves symbols that are not *all* nonformal? Although the latter interpretation is open to fewer objections, it is certainly not what Fodor has in mind in some of his writings, where he contends that symbols have to be *all* formal (Fodor 1980). This would mean that the FSM account entails methodological solipsism, which would imply the absurd conclusion that there cannot be any nonformal symbols in computers, including the very symbols that I am typing in at the moment. All should then be as "blahgrahad," which is just an arbitrary shape or a form of a string of letters. Since Fodor is no longer a method-ological solipsist, I don't think he would accept this proposition today. Therefore, I interpret the FSM thesis as asserting that *all* symbols implicated in computation need *not* be nonformal but *some* may, which is a much weaker claim.

What are symbols on the FSM account? There are several systematically confused meanings of "symbol" in cognitive science. Instead of reviewing them (but see chapter 4, section 1), I suggest that the FSM claim relies on the notion of symbol as used in computability theory. This is how the concept functions in Cutland's introduction of Turing machine M:

At any given time each square of the tape is either blank or contains a single symbol from a fixed finite list of symbols $s_1, s_2, \ldots, s_n$, the *alphabet of M*. (Cutland 1980, 53)

This interpretation of "formal symbol" is unproblematic and shows that Cantwell Smith's claim that the notion always presupposes intentionality is simply false. He writes:

[FSM] explicitly characterizes computing in terms of a semantic or intentional aspect, if for no other reason than that without some such intentional character there would be no warrant in calling it symbol manipulation. (Cantwell Smith 2002, 29)

This claim is based on a questionable way of understanding formality as framed in terms of a rather troublesome theory of naturalized semantics: the difficulties of this theory are taken to be inherited by the notion of the formal symbol. (I criticize a similar theory—as based on a notion of "natural meaning"—in chapter 4, section 2.) But symbols used by the Turing machine may not display any intentionality whatsoever! I agree that if they had to be intentional, inexorable problems would ensue, but the difficulties have mostly been created by Cantwell Smith's idea that, on the FSM account, symbols in cognitive systems are both nonsemantic (formal) and semantic (intentional).[2] That would have been self-contradictory, no doubt; alas, the reductio relies on an admittedly extreme interpretation of the FSM view—even if some authors argue that Fodor's methodological solipsism implies such a position (Devitt 1991).

It transpires immediately that the "formal symbol" in the Turing machine is just a syntactical feature. (It might, though need not, be semantic as well, if some further conditions hold; see chapter 4, section 3 on representational mechanisms.) But unless supplemented with additional constraints, an account of implementation based on syntactical assignments fares no better than the simple-mapping theory; all it says is that physical realizers have to be individuated by virtue of their form (so type-identical states would count as syntactically the same). However, the notion of form still remains quite broad, and there is nothing to block Putnamesque tricks such as defining forms by appeal to clocks (note that phonological information definitely has some temporal form). Indeed, the syntactical view of implementation bears such a close resemblance to the simple-mapping theory that, it stands to reason, they differ but in name. It should come as no surprise, then, that John Searle has famously objected:

If computation is defined in terms of the assignment of syntax, then everything would be a digital computer, because any object whatever could have syntactical ascriptions made to it. You could describe anything in terms of 0's and 1's. . . . For any program and any sufficiently complex object, there is some description of the object under which it is implementing the program. Thus for example the wall behind my back is right now implementing the Wordstar program, because there is some pattern of molecule movements that is isomorphic with the formal structure of Wordstar. (Searle 1992, 207–208)

Fodor's treatment is richer, though. Note, by the way, that he is concerned with computational states ascribable to organisms (Fodor 1975, 75), so his discussion is not intended to cover all kinds of computers (and it does not, as I will show presently). The first condition he sets forth is that these states be directly explicable as relations between an organism and the formulae of the language of thought. In other words, in order to be in a computational state, an organism has to stand in a certain relation to some formula of the language of thought. Second, the class of basic, theoretically relevant relations between the organisms and the formulae should be "pretty small": "Small compared to the class of theoretically relevant relations between the organism and propositions" (ibid., 75). Third, it should be nomologically necessary that for any propositional attitude of the organism (e.g., fearing, learning) there will be a corresponding relation between the organism and the formula(e) such that the organism has the attitude whenever it is in that relation.

These conditions sound promising, as some of the Putnamesque tricks must go against the principle of parsimony, and nomological necessity is anything but a result of an arbitrary description, but they do not apply directly to artifacts, which may not have any propositional attitudes to speak of (I will return to this issue shortly as well). This drawback is obviously owing to the fact that Fodor developed his account of computation strictly for the needs of his theory of explanation in cognitive science, which is a move that I criticized at the outset; but this can be easily remedied. While the first constraint is merely a restatement of the simple-mapping account, the second one (namely, that computational states be ascribed just in case the class of relations between the syntactical symbols and the physical states of the computer is small) appears to undermine Searle's contention that the wall can implement Wordstar. After all, there is no reason to suppose that the class of relevant relations for the wall's implementation of Wordstar will be small; there might be as many of them as there are computational states in the program.

The problem, obviously, is that to use this principle, we would have to translate it into a language applicable to standard computers and not just organisms. So instead of propositions, let us use states of computation. Fodor's principle would therefore require that there be explanatory and predictive purchase in ascribing computation to a physical system; the class of physical states is probably larger than the set of relevant computational states. In this form, this condition seems reasonable; but I set aside this issue as an epistemological matter, which is the focus of the next chapter (see section 2, where I tackle the difference between redescribing the data and finding general principles in them). Suffice it to say that his idea seems to exclude some spurious computational descriptions because they are not methodologically kosher. The reason is that they do not seem to be useful in simplifying the complex physical description or, to put it slightly differently, in compressing physical information.

The third condition is similar to Copeland's requirement that the description support counterfactuals, but it is framed in terms of nomological necessity. In other words, formal symbols have not carried us much further than the simple-mapping theory. The only new thing is the simplicity, or parsimony constraint, which is hard to paraphrase properly for all kinds of computers. The third condition would be more appropriate, were Fodor to say that the necessity should be only nomological and not just logical. (This way the above trick of quantifying over all possible states to achieve counterfactually true descriptions would be stopped; as it stands, nomological necessity, as implied by logical necessity, is satisfied by my Putnamesque counterfactuals.)

Unfortunately, Fodor supplies us with no justification of the claim that computers (as opposed to the planets) need to represent the rules that govern their behavior. Consigned to a footnote, it is not on his list of conditions—but let me try to read it charitably. Plausibly, what Fodor is saying is that computers need to execute programs to be computers, and, for that reason, they need to represent programs to themselves. This is a popular suggestion among philosophers (see, e.g., Cummins 1983). But despite its popularity, it is quite obviously misleading. There are computational devices that do not execute programs; a trivial example is a logical gate, which is a simple circuit implementing a logical operation. Similarly, connectionist networks are trained rather than programmed, and early computers, such as ENIAC, were manually rewired.[3] For some, the principal motivation behind the programmability requirement is that computers should always be universal machines. Sometimes, however, executing programs does not imply being a universal machine—consider jukeboxes or automatic looms (Piccinini 2007a, 517). They cannot simulate every Turing machine (though, contrary to Piccinini's claims, they *are* somewhat trivial computers on his account: like computers, they use input digits to immediately control effectors whose states might be taken to be the output digits).

Moreover, it is not at all clear how to make sense of the traditional program/computer or software/hardware distinction with regard to real-life computers, which may contain multiple levels of software (including software that is hardwired in the device's ROM), so using the notion of software or program to elucidate the concept of realizing a computation might falsely suggest that this distinction is set in stone and absolute (for a similar point, see Lycan 1987, chapter 4). The traditional philosophical claim that the mind is the software of the brain seems deeply confusing because it is hard to find any deep analogies between a standard von Neumann machine that loads a program to its memory and the human brain. For these reasons, I submit that a correct account of implementation should not feature a condition to the effect that computation necessarily involves executing a program.

Where does this leave us? Fodor's suggestion that planets do not compute their orbits because they do not represent Kepler's laws turns out to be unilluminating. Logical gates do not represent the rules of logic either—they simply follow those rules. We may only say that if planetary systems are computers, then they definitely are not software-driven computers, but that is not an especially exciting result. The FSM account, instead of helping with problems that Searle and Putnam discovered, appears to be busy with red herrings.

So maybe the problem with the FSM view is that it is restricted to formal symbols? A related understanding of computation, *qua* semantic notion, is sometimes ascribed to Pylyshyn and Fodor, and prominent defenders of this idea are Brian Cantwell Smith and Robert Cummins (1983). Fodor did write "no computation without representation" (Fodor 1975, 34), but the notion of representation is also deeply troubling in this context. The argument that Fodor offers in this connection is that computation presupposes a medium for representing the structures over which the computational operations are defined.[4]

But the proposition that representations in computational systems always *refer* to something (or have any other discernibly semantic property) is hardly plausible, for if they do, then the symbol-grounding problem (Harnad 1990) makes no sense—computers simply cannot use symbols that do not refer. Alas, I have not found a single argument that computation must operate on nonempty representations, and counterexamples abound (it is not enough to show that there is a representation in a certain computational system; one has to show that they occur in all systems, which is a very strong claim). It is trivial to point to programs or structures that seem completely devoid of reference or meaning, even more than my "blahgrahad." Take this piece of Pascal code:

```
program devoid_of_meaning;
begin
end.
```

It is a correct specification of a program that does nothing.[5] One may also think of a simple connectionist network with two nodes that are interconnected, and the weight is set up so that one node gets activated any time the other activates; it seems to perform a computation—albeit a trivial one (an identity function). What was the referent, again?

Yet Cantwell Smith (2002) insists we need to start by building a theory of semantics to understand computation. However, if there are no symbols that refer (and no programs that are represented to the computer), then it is hard to make anything meaningful out of this suggestion. If you think that representation need not refer (or have any other discernibly semantic property) to be representational, then your use of representation is deflationary; it is nothing but a formal symbol as explicated in this section. Note also that if computation is manipulation or transformation of representations, then would a dog eating a newspaper count as performing a computation? In this example, there is printed linguistic representation, and it is definitely transformed. The account, if it is to be helpful, needs to include at least some of the conditions that we have already outlined—that there is some correspondence between the physical processes and the computational description, that the description supports counterfactuals, and that the correspondence is somehow honest.

So for the sake of argument, let me assume that there is some unproblematic notion of semantic properties involved in computation and that the semantic account of computation includes the above conditions as well. Would that help us with Searle's wall? Not at all. Searle can ascribe some meaning to the wall parts. It will not be intrinsically meaningful but, in his opinion, computers do not enjoy intrinsic intentionality anyway. So appeal to representation is just a dead end.

All these critical remarks notwithstanding, I think there is some rational core to the idea that computation has to do with symbols and representations. As they do not need to enjoy semantic features at all, and remain purely syntactic (*contra* Cantwell Smith), the best way to talk about them is in terms of information that is quantitatively conceived.

4 Information Processing and Computation

In most cognitive science research, the terms "information processing" and "computation" are used interchangeably (e.g., Marr 1982; Pylyshyn 1989;

Floridi 2008a; Boden 2008; Bermúdez 2010). Information-processing systems are systems capable of computation in the broadest sense (*pace* Piccinini and Scarantino 2011). A computational process is one that transforms a stream of input information to produce a stream of information at the output. During the transformation, the process may also rely on information that is part of the selfsame process (internal states of the computational process).

However, the concept of information needs further elucidation. Were I to use the semantic notion of information, my theory would collapse to the version of a semantic account of implementation. A nonsemantic notion is therefore what I have in mind; it might be tempting to call it quantitative but with a caveat that one should not confuse different ways to measure the amount of information with the notion of information itself (MacKay 1969). The most influential theory of information is attributable to Shannon (1948), but it may easily lead to some confusion. For example, David Chalmers (1996b) uses the notion of information, understood generically along the lines of Bateson (1987, 315), as "difference that makes a difference" (Chalmers 1996b, 281), but he calls this notion "Shannon's notion of information," (ibid., 282–283) which is not exactly accurate; Shannon's information theory is probabilistic and involves measurement of the amount of information as communicated in the information channel (see Harms 1998, 480). But that does not mean Bateson's dictum is incorrect; the complete passage defines the "bit" of information, and this notion might be understood in two ways. First, in Shannon's way, which is not what Bateson has in mind, as it requires specifying probabilities or uncertainties of the receiver as to what it will receive in the channel; in the second way it may be understood as a structural notion equivalent to MacKay's logon, which is close to today's computer talk.

MacKay stresses that "it is essential . . . to distinguish the qualitative concept of *information* from the various quantitative measures of *amount-of-information* of which Shannon's is one" (MacKay 1969, 80). He then goes on to distinguish three different measures: (1) Shannon's, which he calls "selective-information-content" (ibid., 80), (2) "structural-information-content" or "logon-content" (ibid., 81), and (3) "metrical-information-content" (ibid.). Shannon's measure indicates the minimum equivalent number of binary steps by which the element concerned may be selected from an ensemble of possible elements.[6] Note that there must be at least two possibilities: if there is just one element in the code, then the receiver's uncertainty as to what element will be communicated is zero, and the probability of this element's occurrence would be unity. Such code would bear no information. If the ensemble in question were just a singleton, there would be no difference to make, so there would be no structural information either. Structural information is understood as the minimum equivalent number of independent features that must be specified—the number of degrees of freedom (independently variable properties) or logical dimensionality of elements. It is measured straightforwardly in logons or the numbers of degrees of freedom. The third notion, of metrical information, indicates the equivalent number of units of evidence that the information provides (e.g., the number of minimally significant events in case of scientific symbolization of an experimental result). These ways to measure the amount of information may be complemented with (4) the Algorithmic Theory of Information (Chaitin 1990); the information content of a binary string S is defined as the size (in bits) of the smallest program required for a canonical universal computer to calculate S.

It is important to note that these concepts of information may be differently applied to the same physical system. Even classical communication theory may be employed in a variety of ways to analyze the same physical process by construing different parameters of the system as parts of the information channel (Barwise and Seligman [1997] stress that this analysis is relative to theoretical interests). The probabilistic notion, it turns out, is especially confusing. Can we analyze mathematical proofs or algorithms as dealing with information? Piccinini and Scarantino (2011) assert that an effective algorithm cannot bear any information because the probability of the event that it halts equals one. A similar point is made against Dretske's use of Shannon's theory by Barwise and Seligman:

Consider, for example, Euclid's theorem that there are infinitely many prime numbers. If it makes any sense to talk of the probability of mathematical statements, then the probability of Euclid's theorem must be 1. Given any prior knowledge k, the conditional probability of Euclid's theorem given k is also 1 and so no signal can carry it as information. This is a serious defect that threatens any probabilistic theory of information flow. (Barwise and Seligman 1997, 17)

But is that really a defect? It is quite obvious that, given knowledge that this is a proof, the probability that it obtains is one. The same consideration would apply to any logical truth, which is precisely why it is called a tautology—it is supposed to produce no new knowledge. The air of paradox disappears once we have realized that, on this analysis, it is assumed that the receiver already knows Euclid's theorem to be true. Indeed, knowing that, the receiver will gain no information owing to the transmission, as the uncertainty as to which of the ensemble of possibilities gets selected is zero. Note, however, that the point of producing proofs is that their receivers have yet to be convinced of their soundness and validity and, consequently, the truth of the conclusion. Also, because the receiver may be an ordinary physical process rather than a cognitive system, it may not be in a position to know anything, including mathematical truths. In such a case, the information is transmitted if the system cannot determine what it will receive. So the text of the proof, say, in Morse code, sent to a system that can print dots and dashes but does not know anything about their mathematical interpretation, *would* be information for that system.

Similarly, if the process that receives the output data from an algorithmic process already knows the data, or if its uncertainty as to what it will receive from the algorithm is zero, then the algorithm does not transmit information. Those who think that algorithms cannot process information because they are deterministic might be tempted to say that simply introducing random noise to the algorithm would make it informative. But this is absurd. It is as if the communication theorist enhanced the channel's information-bearing capacities by adding random noise. Telecommunications engineers do not wake up screaming in the middle of the night, haunted by the nightmare of perfect, noiseless channels. Such channels do not make the content noninformative. (Note that Shannon [1948] proved that adding appropriate levels of redundancy to the code can enhance the channel's reliability and cancel out the effects of noise, which would be self-defeating if less random noise meant less information.) The moral is this: Always specify the receiver when performing Shannon's probabilistic analysis; otherwise, (apparently) counterintuitive results will be obtained.

Another important point about Shannon's theory is that it defines the average amount of information in the channel. To determine this average measure, however, an immense number of data may be required, which might be hard to obtain in the case of natural systems. As Rolls and Treves say, "With real neurophysiological data, because we typically have limited numbers of trials, it is difficult from the frequency of each possible neuro-nal response to accurately estimate its probability" (2011, 452). This is the limited sampling problem:

If the responses are continuous quantities, the probability of observing exactly the same response twice is infinitesimal. In the absence of further manipulation, this would imply that each stimulus generates its own set of unique responses, therefore any response that has actually occurred could be associated unequivocally with one stimulus, and the mutual information would always equal the entropy of the stimulus set. This absurdity shows that in order to estimate probability densities from experimental frequencies, one has to resort to some regularizing manipulation. (Rolls and Treves 2011, 452)

For my purposes, any of measure of information may be applied to analyze information-processing systems; the input stream of information is a sequence of states of the vehicles (particulars classified under some type) as distinguished by the process itself. In other words, there is some structural information as long as the vehicle has at least one degree of freedom. (Metrical information content is less relevant here, but it may be applied to analyze the reliability of the physical bearer to carry informa-tion.) Also, there is some Shannon information as long as the receiver of input information cannot determine with absolute certainty the probabil-ity of the vehicle's being in a particular state. It is almost the same require-ment: the vehicle needs to have at least one degree of freedom or be able to assume at least two distinct states, *and* the receiver may not determine in advance which condition will apply. So computation is simply informa-tion processing:

Computation is a kind of flow of information, in at least the most general sense of the latter idea: any computation takes as input some distribution of possibilities in a state space and determines, by physical processes, some other distribution consis-tent with that input. (Ladyman et al. 2007, 208)

Some people may be tempted to think that information flows if and only if it is preserved; for example, if it is encoded in terms of propositions, their truth-values cannot be altered. But all it takes for information to flow is the presence of some stable relationships in the physical system. In particular, some structural-information-content may be lost during the flow in an algorithmic process. If an algorithm that is able to process any kind of binary-encoded natural number (of length less than n bits) deter-mines that 1011 is an odd number by saying "1," then you cannot recover the sequence of input bits even if you know the output of the algorithm, as one bit cannot encode all possible natural numbers (of length greater than one but less than n). Similarly, information need not be veridical to be processed or to flow; a computational process may freely negate a true proposition and still count as information processing. The only correct way to describe this flow as truth-preserving in logic is to use special, unsound logics (as Barwise and Seligman [1997] actually do).

This characterization of computation in terms of information processing is very broad, and it must be elaborated on, especially when applied to physical cognitive systems. The first obvious observation one can make in this connection is that some information may be digital. That is, there is a finite, denumerable set of states of the vehicle of information recognizable by the computational process in question, and the output of the process is similarly couched in terms of that set. In computability theory, this set of states is called the "alphabet," and the distinct states that are recognized are called "the symbols defined over the alphabet" (as in Cutland 1980). The very notion of "symbol" may lead to confusion; when realized physically, a symbol is just a state of the vehicle that is reliably distinguished by the physical process from the rest of the states. The symbol, therefore, is not presupposed to carry any semantic information whatsoever. It is just one of the various members of a set.

In the case of analog processing, the range of values that the vehicle may assume is restricted, but continuous—that is, infinite (in the limit). Standard analog processing, as known in the history of computing (e.g., in Soviet attempts to create universal analog computers in the 1950s), does not rely on any infinite range of values; the measurement operations of physical devices are always of limited resolution, so the range is never actually continuous. Due to imperfect measurement, even a slide rule, which is regarded as a simple analog device, operates on discrete values. For this reason, one may even be tempted to call a slide rule digital, even if its operation does not always include the same number of steps: the operation of the slide rule is as continuous as the range of values it processes. However, there are (hypothetical) hypercomputational analog computers that are designed to exploit an actual infinity of values (see, e.g., Siegelmann 1994), so one should not dismiss this possibility a priori by imposing constraints on the range of values (even if one believes the prospects of physically implementing such computers to be dim). An interesting point is that hypercomputational devices, if they react differently to all of the infinite range of values, would process an infinite amount of information (in terms of both structural and selective information content). Again, such devices need perfect measurement, but they are not obviously physically impossible (Copeland 2004, 258). (Also, note that actual infinity is just a sufficient condition of hypercomputation—not a necessary one.)

An information-processing system may start processing with any input. Although inputless information-processing systems might be rare, nothing prevents us from saying that they process information if only they have any information-bearing states as their internal parts. It is required, however, that the process have no empty information-bearing output states. The input and output information states can form streams; that is, they may be accepted by the process as a sequence of input elements. It is important that the input and output streams are ordered—that is, the chronological and spatial ordering of the code tokens matters. The initial state of a computational process (i.e., one that takes in the input information stream, if any) has to be causally related to the final state of the process. In other words, a physical implementation of a computation is a causal process. Note that cyclic information processing, as found in interactive computational processes that abound in standard routines that support user interfaces, may take few bits of information on input to start a relatively complex process that relies on some other information source. An obvious example is when a single keystroke is used to activate a program that reads a file into the memory and processes it in some way. Similarly, there might be processes in the neural system that are cyclic and use simple inputs to control further processing rather than to process those very inputs.

One could object that this characterization of computation in terms of information processing is actually a variant of a semantic account of computation. I think this is in error. Just like Fresco (2010) and Piccinini (2006), I suggest that information processed by a computer need not refer to, or be about, anything in order to be the inputs or outputs of a computation. Admittedly, physically realized input and output states trivially carry semantic information as parts of the causal nexus (effects are indicative of their causes; see Collier 1999). But they need not be about any *distal* events—they are just informative of whatever caused them. So they need not be representational in the normal sense presupposed in cognitive science; no properly semantic characterization of computation is required for the theory of computation. (For this reason, the information-processing account is closer to FSM than to the semantic theory.) Nevertheless, most information-processing systems considered in cognitive science do deal with semantic information in its proper sense. I will expand on this subject in chapter 4.

The above account of computation does not presuppose any model of computation from computability theory. (Note that I use "model of computation" in the technical sense here, as it is used to describe Markov algorithms, lambda calculus, or Turing machines.) It is not up to philosophers to decide which such model is correct; this is a question for mathematics and computer science. Although there is a paradigm case of computation—the classical Turing-Church digital computation—we leave open the possibility that there are physical processes capable of computing nonrecursive, hyper-Turing functions. My account is a brand of *transparent computationalism*, as defined by Ron Chrisley (2000): it can accommodate any kind of computation, provided the computation is specified in terms of relations between input and output information states. It seems that digital von Neumann machines, membrane analog computers, quantum computers, and perceptrons all come out as computational in this view.

One of the reasons why one mathematical understanding of computation should not be assumed as the only possible physical implementation of computational processes is that, in real information-processing systems, some parts may operate according to different principles than others. To take a recent example, O'Reilly argues that the prefrontal cortex (PFC) implements discrete, digital computation—exhibiting such behaviors as bistability and gating, which are specific for digital machines—but the rest of the cortex may be a fundamentally analog system operating on graded, distributed information (O'Reilly 2006). It can also be argued that relatively classical explanations in computational neuroscience, such as Marr's (1982) theory of edge detection, presupposed that the brain performs analog computations (Shagrir 2010c,d). There are also hypotheses that neural computation is neither completely digital, nor completely analog (Piccinini and Bahar, forthcoming). A scientifically informed account of implementation should not render such claims a priori falsehoods!

The account of computation in terms of information processing may be considered to be a refined version of the FSM view—one that does not have problems with the notions of symbols or formality as they are unpacked in terms of information (and Pylyshyn does analyze cognition in terms of "natural information-processing," which he equates with computation; see Pylyshyn 1989, 85). At the same time it is more general, as it accommodates hypercomputation, whereas the classical formulations of the FSM view seemed to presuppose the Church-Turing thesis. However, generality comes at a price: it might be objected that the account is so broad that it makes the notion in question trivial. This is why we need to introduce further distinctions. First of all, a complete specification of a computational process has to include an exact description of the mathematical model of computation involved. Digital computation, for example, requires that there be a finite set of states recognized by the system; effective digital computation has to be specified in terms of partial recursive functions or something equivalent to them (such as universal Turing machines). Second, there are problems that an identification of computing with information processing does not address at all. For instance, the initial and final states of information-bearing vehicles are causally linked; otherwise, there would be no flow of information. This means, among other things, that it is possible, in principle, to extend the boundaries of the computational system ad infinitum because, plausibly, (almost) every cause has at least one cause of its own, and (almost) every effect has at least one effect. With every causal chain stretching as far back as the Big Bang, we do not know, at this stage, how to delineate a system implementing a computation. By saying that computation is information processing, we only point out what the system does and not what the system is. This will need to be tackled.

Another difficulty is that without an account of the organization of computational systems, it is not at all clear what would be the proper mapping of a mathematical model to the physical process. For example, must the initial and final states be at the same level of organization? It could be objected that these mappings are usually arbitrary and observer-dependent. Moreover, one cannot distinguish a system that is merely modeled computationally (such as a pail of water, a tornado, Searle's wall, the Solar system, or the interactions of gas molecules) from a system that actually does compute, such as a desktop computer (see Piccinini 2007b). Even though the scope of computational modeling might be universal (one can model any empirical phenomenon with arbitrary precision even in a digital fashion, given sufficient computational resources), it seems important to account for the information-processing dynamics of certain systems. So far, then, our theory of implementation formulated in terms of information processing fares no better than its predecessors, as it is hard to envisage a physical system in space-time that would not fulfill the simple requirement of having at least two states that are causally linked.[7]

Some writers argue that this is all we might ever get from an account of implementation (Shagrir 2010b,c). I agree with Shagrir that there are theories in cognitive science that may be understood in this way. But there is a problem with defining the boundaries of the information processors. The theory of implementation should say what kinds of entities implemented computers are; this is an ontological component of the computational explanation. It is not to replace (or be confused with) the mathematical theory of computation and the epistemological theory of computational explanation.

Shortly, we want to know what kind of ontological commitments a computational theory of mind or a theory of biological information processing (e.g., a theory of how information is processed when amino acids are synthesized) could have. Theories indicating that these are entities described with computational notions seem to change the subject; such theories are as uninformative as theories of quantum collapse that say it is a process described in quantum mechanics. The latter is undoubtedly true, but we want to know what it is about some physical processes or events that calls for notion of collapse. Similarly, a deflationary account of computation does not answer the question of *why* we would want to apply computational notions to physical reality. The information-processing story answers that in part, but it still cannot individuate the computational systems in question. Where do they start and end? How to talk of their parts? To answer these kinds of questions, we need to add further constraints to the information-processing account, which is the job of the next section.

5 Computational Mechanisms

My position on the nature of computational processes appeals to functional and structural notions such as "mechanism" and "level of organization." Computations are performed by mechanisms, which is to say that computational functions are realized by the organization of the mechanism's parts. Though explicated in a variety of ways (see, e.g., Bechtel 2008a; Machamer, Darden, and Craver 2000; Craver 2007; Glennan 1996, 2002), the concept of mechanism plays a major role in contemporary philosophy of science. An account of computation that relies on the concept of mechanism is due to Gualtiero Piccinini (2007a). Some earlier, similar theories were cast in functional or teleological terms (Lycan 1987; Sterelny 1990); another related conception of implementation, spelled out in causal-structural terms, is attributable to David Chalmers (2011).

There are several advantages to using the notion of mechanism to explain the realization of a computation. First, in order to talk about a mechanism, one needs to distinguish its parts and their interactions from the environment; this helps to solve the problem of delineating the boundaries of computational systems by pointing to what is constitutively relevant for them (see Craver 2007, 139–160). Second, mechanisms are used in the context of mechanistic explanation, which in turn is a type of causal explanation. (I will return to the explanatory functions of mechanisms in the following chapter.) Carl Craver (2007) argues convincingly that, in this case, causation should be construed along interventionist lines (see Woodward 2003 and Pearl 2000). This helps to show that computational systems must support counterfactual conditionals and that their structure is to be described by structural equations that take into account interventions (natural and man-made). Third, mechanisms in the special sciences are realized by structures comprising multiple levels of organization—one of which is the computational level. In light of this mechanistic analysis, it is obvious that the computational level cannot be the only one in the purported mechanism. In other words, the lowest level must be realized by its constitutive parts, or "bottom out" (to use the phrase from Machamer, Darden, and Craver 2000). Even if the lowest level may be described computationally itself (e.g., as a collection of computational submechanisms), it needs to have a different description than the whole mechanism, as I will argue in this section. This is to say that computations occur only in highly organized entities and cannot float freely without any realizing machinery.

These advantages help to handle the counterexamples to the simple-mapping account that were devised by Searle and Putnam. According to the mechanistic view, we have to match a complete trajectory of states in

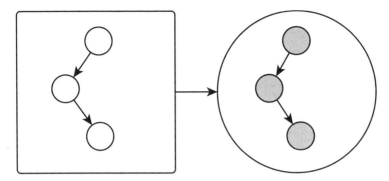

Figure 2.5
The complete trajectory of physical states (left) should correspond to the model of computation (right).

the computational model to the causal pathway in the physical process, and that is much harder to do than describing a random sequence of causes and effects as a sequence of states of the computation (see figure 2.5). The mechanistic approach naturally includes the constraints proposed both in counterfactual theories of implementation (e.g., Copeland 1996) and in its causal accounts (Chalmers 2011).

Computational mechanisms bottom out in noncomputational structures. This requires that higher-level structures be identified on the lower level of the mechanism's composition, and as such, it is considered a good strategy for avoiding charges of triviality that were voiced against functionalism—in particular against the functionalist simple-mapping account (Godfrey-Smith 2008). This identification simply disallows arbitrary ascriptions that are not grounded in lower levels. To show how bottoming out helps against Wordstar walls, I need to introduce more detail.

To say that a mechanistically adequate model of computation is implemented is to say that the input and output information streams are causally linked and that this link, along with the specific structure of information processing, is completely specified. If the link is not completely (and truly) described, it is possible that the flow of information is disrupted and that the information processing has some additional features that influence the overall result of the computation. The description of a mechanistically adequate model of computation usually comprises two parts: (1) An abstract specification of computation, which should include all the variables causally relevant for processing, and (2) a complete blue-print of the mechanism on three levels of its organization. In mechanistic explanation, there are no mechanisms as such; there are only mechanisms *of* something, and here that something is (1). By providing the blueprint of the system, we explain its capacity, or competence, which is abstractly specified in (1).

For example, the adequate model of a connectionist network realized by a dedicated chip will include (1) the specification of the network in terms of its connections, weights, and other computationally relevant parameters, so that the exact pathway of processing is known, and (2) the blueprint of the hardware, along with its speed characteristics, materials, and so forth. Theoretically, one could have a single description that would serve both purposes; in practice, it is easier for us to separate them and use both parts for different purposes. For example, one could build a software simulation of the network by focusing only on its computational function. Notice, however, that many extant computational descriptions in cognitive science are—exactly for this reason—only mechanism schemas or mere abstract specifications of their computational roles (Piccinini and Craver 2011).

Note that the requirement to adequately spell the complete model of the mechanism may seem to be quite liberal. Say we have a complete description of the mechanism of a particular mousetrap. Couldn't we redescribe its causal model in computational terms simply by saying that the mousetrap is a computer that processes the input information given in terms of a living mouse and yields an output as a negation realized as a dead mouse? Obviously, one could. But the requirement that the description be complete is not a sufficient condition of being a computer—it is only necessary. To decide whether the mousetrap is a computer or not, we need to see how it is embedded in the environment and what role it plays. Actually, one could build a computational system containing a specific trap that is a "mouse sensor" and use it to send a text message to the janitor of a grain silo. We do not want to exclude this case a priori!

Mechanisms are multilevel systems: the behavioral capacity of the whole system, though generated by the operations of its components, is usually different from the behavior of its parts. The notion of a level of organization is known to be especially hard to explicate (see, e.g., Wimsatt 2007 for an incomplete list of over two dozen criteria that levels of organization must satisfy to be real levels), but we are now interested only in levels of composition (Craver 2007, chapter 5). Every computational mechanism has at least three levels of organization: (1) A *constitutive* (−1) level, which is the lowest level in the given analysis; (2) an *isolated* (0) level, at which the parts of the mechanism are specified along with their interactions (activities or operations); and (3) the *contextual* (+1) level, at which the function the mechanism performs is seen in a broader context (e.g., the context for an embedded computer might be a story about how it controls a missile). For cognitive systems, the contextual level includes general cognitive mechanisms and the environment of the system; the isolated level features the computational processes that contribute to cognitive processing, which match the abstract specification of the computation and its causal model. At the constitutive level, the structures that realize the computation are described.

Computational organization is itself realized by lower-level mechanisms, which are not computational—at least, they are not computational in the same way. So while the constitutive level might comprise information-processing submechanisms, the lower-level mechanisms would be causally contributing to the isolated level computational function, and as such, their nature—computational or not—is irrelevant. Lower-level mechanisms are not computational in the sense that they constitute the causal structure that matches the computational characteristics of the whole mechanism. A connectionist network could be realized immediately by dedicated electronic submechanisms, or as software saved in an EPROM of a chip that is itself realized by some electronic machinery, or even by some hybrid electronic-biological hardware. The only requirement is that we understand in what way the constitutive level contributes to the functioning of the isolated level.

Everything below the constitutive level usually falls outside the purview of a description of the computational mechanism (so, e.g., a story about how a game console works need not engage quantum mechanics); the isolated level is usually described in a fairly abstract fashion. The question of which levels of organization a given description should include cannot be answered without reference to the explanatory goals of a particular research project; I defer this discussion to chapter 3.

Note that only the isolated level corresponds to what was the traditional focus of philosophical theories of implementation: it is the only level at which the abstract specification of computation is to reflect causal organization. The causal organization of a mechanism will inevitably be more complex than any abstract specification; yet, since we are building a constitutive explanation of a given capacity, our model will include only those causal variables that are relevant to that capacity. In normal conditions, the causal model of a mechanism will specify all the dynamics that reflect all the valid state transitions of the abstract specification of computation. However, it will also contain a specification of interventions that will render the computational mechanism malfunctional.

The isolated level is not autonomous, and the parts of the mechanism at this level should be localized, or identified, at the lower level independently of their computational roles (see figure 2.6)—otherwise, the whole description of the isolated level will remain unjustified. Notice that this principle of bottoming out (Machamer, Darden, and Craver 2000) does not preclude the possibility of some parts being constituted in a distributed manner; however, since disjunctive states at the isolated level will have no counterparts at the lower level, it does block Putnam-style tricks. Description of the lower level does not contain these disjunctions at all. Obviously, one could map a higher-level disjunction to a lower-level disjunction, but this does not satisfy the bottoming-out requirement. The principles of identity at the constitutive level come only from this level. You cannot explain how higher-level objects are constituted at the lower level by

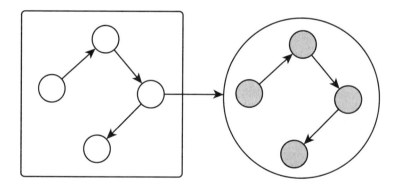

Figure 2.6
The computational-level entities should be identified on the constitutive level. Here, component 3 in the causal chain (left) is identified with a certain four-part subsystem (right).

saying that their constitution is determined at the higher level. This kind of top-down determination will fly in the face of the bottoming-out principle (for a similar consideration, see Piccinini 2010). Note also that this is not incompatible with using higher-level entities heuristically to discover lower-level entities. (For a story about how the psychological theory of memory contributed to our understanding of the hippocampus, see Craver 2007).

So instead of adding special provisos against using nonnatural mappings, we simply require that the computational description at the isolated level not be completely independent of the constitutive level. In a similar vein, mechanistic explanation offers a way to individuate mechanisms (Craver 2007) that allows us to distinguish real systems from a mere hodge-podge of physical states. More importantly, parts of mechanisms are functional, as I elaborate in section 5.1, and that helps to exclude from the set of mechanisms some physical systems—such as tornadoes, piles of sand, or planetary systems (Piccinini 2007b, 2010)—that normally do not realize any information-processing capacity.

Notice also that there are no natural boundaries to be found that correspond to the *computing* machinery that realizes Wordstar on the wall. The wall is not organized as an information-processing mechanism; there is no constitutive level that corresponds to the isolated level that does the computing, as there is no physically based decomposition of the word-processing system. This is why the wall "computer," in contrast to a normal wall, does not have proper individuation criteria: its constitutive level does not allow one to carve the system in the same way as the computational description. In short, there are no lower-level regularities that enable us to identify the parts of the mechanism in the wall *independently* of the computational role that they are supposed to play. The Wordstar computer needs to have a CPU, for example, so there needs to be a CPU at the constitutive level of the wall. The problem is that the causal model of the CPU is quite specific, and it requires the activity of intricately orchestrated component parts. Such a causal structure, for all I know about standard walls, is simply not to be found. We could call some parts of the wall "CPU," but that does not mean that this statement would be by itself true! It would be true only if one could find specific causal structures of the CPU that both fulfill all the normative constraints of the interventionist theory of causation and are empirically validated. These constraints simply disallow arbitrary groupings of physical entities that are at the core of Putnamesque examples. Mere redescription of parts of the wall in terms of the CPU does not make them causally structured in the required way. They need to really operate this way (see section 5.1).

The wall, moreover, seems to be an aggregate of parts, so no wonder it is not properly organized to be a computer. Walls, pails of water, or chewing gum lack the multilevel organization required to implement computations; by the very organization of these mechanisms, they do not have independently individuated parts that have functions.

I will elaborate on causal organization and functionality in section 5.1, as these are important aspects of the mechanistic account. Note also that my mechanistic view of implementation adheres to transparent computationalism, and this is how it differs from other causal or mechanistic theories. There is actual danger that physical realization of computation will be conflated with just one kind of mechanism capable of implementing only one kind of computational model. For example, the account of David Chalmers, who claims that computational processes should be modeled as combinatorial state automata (CSA), has exactly this deficit (Chalmers 1996a). Several other structural approaches share this deficiency (see Shagrir 2006). Piccinini in his work (e.g., Piccinini 2007a) discussed only digital computation, and has only recently admitted the need to accommodate nonstandard models under the umbrella of "generic computation" (Piccinini and Scarantino 2011). His general scheme of describing computational models is based on abstract string-rewriting accounts of computation—that is, computation is construed as rewriting strings of digits:

A computing mechanism is a mechanism whose function is to generate output strings from input strings and (possibly) internal states, in accordance with a general rule that applies to all relevant strings and depends on the input strings and (possibly) internal states for its application. (Piccinini 2007a, 501)

This is suitable for digital computation, but not for computing mechanisms that take genuine real numbers as input, and it is not obvious how to make it more general to cover generic computation. Moreover, string-rewriting systems are not so easy to map onto causal relationships as state-transition systems; if states are missing from the description, it is not at all clear what one should take as relata of the causal relations. Obviously, not string values—rather, something that triggers the rewriting operation. If so, however, string rewriting is mapped onto state transitions, and the original model is actually never used directly. In contrast to Piccinini, I think that the mechanistic account should be *directly* linked with causal explanation. The most suitable way to do so is to use the abstract model of computation, related to its causally relevant parameters, and the engineering blueprint of the mechanism on all three levels of its composition. So, for example, for a Turing machine realized for fun using LEGO blocks, the adequate abstract model would be the Turing machine and the description of how it is built from blocks. But for my laptop, a description in terms of the Turing machine is not accurate, as my laptop does not have a tape as its causally relevant part.

Standard models of computation, such as a Turing machine or lambda calculus, are highly abstract and describe only how functions are computed. For example, the Turing machine formalism does not say how much actual time one step of computation takes (note, however, that it is possible to create a formalism that does exactly that; see Nagy and Akl 2011). It is important not to confuse this notion of models of computation with the one used in the philosophy of science to, for example, refer to computational models of the weather. Both are formal models that are supposed to represent reality (accordingly, computation or weather) but with different purposes and in different ways. We usually talk in terms of standard computational models when we specify the abstract structure of computation, as this makes things easier. But note that these models do not include any details of implementation. Any theory of implementation that is satisfied with an abstract model that omits all implementation details is therefore self-defeating. These details are important when validating the models that engender hypotheses about the causal structure of processes. This is why the theory of implementation needs to include them. In short, implementation should never be conflated with the formal model of computation.

Adversaries of the computational theory of mind point out that Turing machines are formal, atemporal structures (e.g., Bickhard and Terveen 1995; Wheeler 2005). They take the computational theory of the mind as implying that only abstract machines are explanatorily important for understanding cognition. It need not—indeed it should not—be the case, but this kind of objection is based on the same conflation of implementation with the formal model and should be eschewed by computationalists!

The organization of mechanisms is temporal and spatial—in contradistinction to the abstract models of computation, physical computer parts have spatiotemporal properties. Besides displaying the behaviors that contribute to their computational role, they also have additional properties (e.g., heat) that may be used to detect them and predict failures of computation. Abstract computationalism would be indeed doomed as a theoretical position.

5.1   Computational Mechanisms Are Causal Systems Organized Functionally

Computational mechanisms may vary considerably. They must satisfy the criteria of the simple-mapping account of computation; that is, there must be at least two states of the mechanism—the initial one and the final one. I understand them in terms of information, as explicated in section 4 above. Information flow is described causally. Causality in this case is stronger than the generic requirement imposed on functional accounts: it is based on an interventionist theory of causation.

The interventionist theory of causation makes specification of the relation between the initial information-bearing state and the final state easier, as in this account, the relata of the causal relation are variables. A necessary and sufficient condition for X to be a direct cause of Y, with respect to some variable set V, is that there be a possible intervention on X such that it will change the value of Y (or the probability distribution of Y) when all the other variables in V—those except X and Y—are held fixed at some value by interventions (Woodward 2003, 55). In other words, there is a possible intervention modifying the value of the initial state (X) that will change the final state (Y). The relationships of causal variables are described using structural equations or directed acyclic graphs (Pearl 2000). The latter especially supply an intuitive way of representing causality (see figure 2.7).

A correct description of a computational mechanism will specify the parts of the mechanism and their organization. The organization described in computational terms is usually called "architecture." Architecture is spelled out in terms of a formally-specified model of computation. Different models of computation rely on different architectures; a mechanism that realizes a Turing machine with two tapes (putting aside the problem that the tape will not be infinite in a finite universe that we might find ourselves in) will be different from a mechanism that realizes a per-

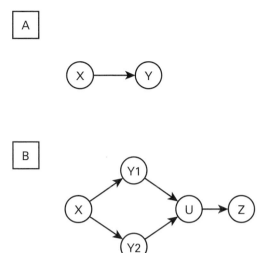

Figure 2.7
Directed acyclic graphs showing causality relationships: (A) X is the cause of Y; (B) X causes Y1 or Y2, both of which then cause U, and U causes Z.

ceptron or a finite-state automaton (FSA), though it might compute the same function. The specification of the organization should be granular enough to distinguish among different internal organizations of computational mechanisms that correspond to different models of computation. For example, two different computational mechanisms may alphabetize a list of words. The input/output relation will be identical for both mechanisms: they both take an unsorted list of words as input and spit out a sorted one as output. Yet the individual steps of computation, which correspond to the particular causally relevant states of the mechanism, will be different, because the parts and operations of the mechanism are individuated in terms of the computation model employed.

Furthermore, Putnam-style tricks become inadmissible as soon we apply the norms of causal explanation offered by the interventionist account (Pearl 2000); the complexity of causal graphs that yield the same predictions might be a reason to discard the baroque disjunctive graphs. More importantly, because interventionist causation requires that causal factors be specified as relevant against the background of other possible factors (so-called contrast classes), arbitrary disjunctions will fail. This means that the causal trajectory that we need to match to the model of computation, even if it is not just out there waiting to be discovered as a theory-independent entity, cannot be replaced with a mere disjunctive hodge-podge of states.

The mechanisms should be specified completely in the appropriate formal terms, as a given model of computation requires (e.g., as a definition of a Turing machine, a Pascal program and its machine interpreter, or a machine code). The level of detail here should be sufficient to reimplement the computation on the same kind of machine. Otherwise, what we have is a mechanism sketch or a specification that contains gaps (Piccinini and Craver 2011). The requirement of providing a complete specification of the model of computation and an engineering blueprint of the machine stems from the fact that only complete computational mechanisms can be physically implemented. If we leave out any detail that is needed to replicate the computational device in full, we have not described the organization realized by the mechanism, which means that different variants are possible. If we do not know some details of how a device operates, a sketch might be inevitable. But it is essential to note that a computational mechanism cannot be partially implemented. Physical implementation is always complete; a computational mechanism will not work otherwise. Functioning mechanisms out there do not have gaps or black boxes, so, from the ontological point of view, implementation requires full description. It is only from an epistemological point of view that computational explanation admits of idealization (e.g., specifying the minimal core of the computation or distorting the complexity of the device to simplify it; see chapter 3, section 1).

The abstract organization that corresponds to the computational architecture specifies its design in a functional way. Let me elaborate. It only specifies the parts of the mechanisms as *types* (as opposed to *tokens*). The role of the mechanism parts is thereby fixed by its internal organization; parts of the mechanism that are not fixed as types by its design are not candidates for strict functional ascriptions (for a detailed analysis, see Krohs 2007; Krohs's account is in many respects similar to that of Wimsatt 2002).[8] In other words, the functional role of a component is one of its causal roles, such that it contributes to the system behavior of the mechanism (as in the classical analytical account in Cummins 1975; for a mechanistic variant of this account, see Craver 2001), but the organization of the mechanism is based on the process of selection of its parts as types.

Note that the requirement of functionality spelled out here does not pertain to all kinds of mechanisms in science. The proponents of the mechanistic theory of implementation of computation do stress that functionality is important (Piccinini 2007b). There is a reason for that: not all systems are functional, and functionality is a phenomenon that is not just a matter of actual organization. It is a property of robustly organized systems that are structured to have some capacities; their capacities are not just coincidences. For example, the molecules of water in a pail do not constitute parts of a functional mechanism whose capacity is to indicate the volume of the pail if there was no selection process that selected those parts as having certain type-level properties. However, if someone uses a certain volume of water to measure the capacity of the pail, then the molecules are functional—and might be replaced by anything else that has a property of maintaining the same volume when poured (so highly evaporative liquids will not do). So a pail of water might be a part of a computational system in such a case! This is not so surprising given that we do know of hydraulic computers (e.g., the MONIAC that was used in 1950s for simulations of a Keynesian economy [Bissell 2007]). But obviously their organization was much more intricate than that of a single pail.

Unless we know the current operation of the mechanism and the selection process that originated it, we cannot determine whether a system is functional or not. A random match between the physical structure and a design is not enough to determine functionality; it suffices only to ascertain that a structure has some disposition. It follows that random matching of physical systems and, say, a given finite-state machine will not suffice for implementation if conceived in this functional way. Note however that having a function is a matter of fact and not just an epistemological consideration or something only in the eye of the beholder. It is just that we cannot determine function by observing only the current organization of the system.

The requirement that the components identified in the architecture correspond to the functional parts is hard for many physical systems to satisfy. For example, a pail of water lacks the functional organization necessary to fix the types of parts that would correspond to some model of computation. Computational artifacts, however, all have designed parts that are fixed as types rather than just tokens; the design of an artifact makes no reference to its parts' history, label, or spacetime location before assembling the mechanism. (Relative spacetime locations of parts are included in the documentation for sure, but they refer to the position of parts in the mechanism; the previous location of parts is irrelevant.) This is why we would find it weird if the documentation for assembling a computer contained instructions referring to the surname of a person who controlled the production process for CPUs, the location of the factory, or a history of the raw materials used for the production of capacitors; we would probably think that these properties must make some difference to more type-level properties of the parts (e.g., because of certain kinds of side-effects in the production process). However, if told that these properties make no difference and that these parts are indistinguishable by engineering standards from parts from other factories, we would think that the documentation is simply wrong or sarcastic.

There are considerable controversies over the notion of function in the philosophy of science (for a representative sample, see Ariew, Cummins, and Perlman 2002). Krohs's (2007) construal, which I endorse, is less popular than its two major competitors—role-function theory and history-function theory—so an elucidation is in order. The role-function, as defined by Cummins (1975), is geared toward an analysis of complex capacities in terms of their subcapacities. According to Cummins, something is functional if it contributes (causally) to some capacity. So, for example, if we identify locomotion as the function or capacity of a bicycle, then the wheels will be among its functional elements (presumably, one cannot ride a bicycle without wheels, since wheels realize the subcapacity of turning) as opposed to, say, a bell mounted on the bicycle's handlebar (one can ride a bicycle without a bell—i.e., ringing is *not* a subcapacity of locomotion). A major problem with this type of analysis is that it cannot distinguish function from malfunction (malfunction may be defined as yet another function). The reason is that there is no single privileged way of determining the system capacity of a mechanism. The human heart, for example, does not only perform the blood-pumping function; it also makes throbbing sounds. Both are perfectly fine as system capacities for functional analysis. But so is the capacity to burst out with blood when shot. What is lacking in functional analysis, therefore, is a method of identifying a system's proper capacities—*the* capacities, as it were.

This is a serious shortcoming because, generally speaking, malfunctions provide researchers with crucial information about mechanisms; miscomputations are of special interest in the case of computational mechanisms (Piccinini 2007b). They are especially informative about the level of implementation, and they can be exploited to discover the causal structure and *design* of the system as displaying some computational capacity.

The theory of history-function is an attempt to introduce the function/ malfunction distinction to functional analysis. The idea behind it is that in order to determine whether an entity has a (proper) function, we need to look at its history (Wright 1973; Millikan 1984, 2002). Has the entity been selected because it works this way and not another? If so, then it is functional. Note that whatever was not selected in the past because of the way it worked is not functional. So bursting-with-blood-when-shot is not a functional capacity; it is a malfunction. So far, so good. However, the capacities of the first instances of entities, without any previous selection history, are not functional at all.

This is not a huge problem in biology, where the first specimens of a species are hard to come by. But in technology, it becomes a major hurdle. ENIAC did not have a predecessor, but its capacity to compute was functional all right; it definitely was not a malfunction! Another difficulty, which is closely related to this one, is that function is a product of a historical chain of events. As such, it cannot be causally relevant. This is especially vivid when we consider a Swampman: my own atom-by-atom copy (Davidson 1987). The Swampman, poor creature, is the outcome of a highly unlikely accident, so he lacks my selection history; he came into being when a swamp nearby was hit by lightning. Funny stories aside, the historical notion of function makes functions causally irrelevant. All of the causally relevant properties of a Swampman, including all organization-related features, are the same as in myself; yet, on the historical account, none of them performs a function. What does this mean? The history-function theory vindicates function by making it epiphenomenal and irrelevant to the functioning of a mechanism. (For a similar argument, see Bickhard 2008; Dennett [1991a] stresses that historical facts about function are "inert.")

There have been proposals to treat role-function and history-function as serving different theoretical purposes and thereby not in competition with each other (Millikan 2002). That might be so; but I want the notion of function to be both causally relevant and useful in distinguishing the functional from the malfunctional. Krohs's (2007) theory seems to satisfy both criteria.[9] The organization of a system, or its design, is causally relevant to its functioning. Also, if the system is no longer organized as designed, it becomes malfunctional. In the case of biological systems, design is best conceived of in genetic terms; in the case of artifacts, design is created by engineers. In other words, this account accommodates both artifacts and natural systems; that is also an advantage. Man-made computers are artifacts, but if the computational theory of mind is true, then brains are also computers—without having been engineered by humans.

The design of a system is the outcome of its history, but its causal relevance does not depend on that. A necessary condition of an entity being part of functional organization is that it is selected as type by the design. Again, this is clear in the genetic case (traits are not selected for as spatiotemporal particulars but as types) and in the context of technology; engineers do not care about the token identity of the parts of functional systems. That functions are ascribed to types explicates the intuition about function—namely that it is a fairly abstract property that may be substrate-neutral (more on that in section 5.2). As long as it is a type, token-level differences are irrelevant to the design.

In short, functions are bound to the design of a mechanism, and computational mechanisms need to have functional parts in order to be mechanisms. Otherwise, they lack the organization that warrants their computational functioning. Also, the organization of a computational mechanism is usually quite complex. Complex mechanisms display behaviors that are significantly different from the behavior of their parts; they are not mere aggregates (Wimsatt 1997). According to Wimsatt, aggregativity requires the fulfillment of all the following conditions:

(W1) *rearrangement* or *intersubstitution* of parts does not change the system behavior;

(W2) *addition* or *subtraction* of parts changes the system behavior only qualitatively;

(W3) system behavior remains invariant under *disaggregation* or *reaggregation*;

(W4) there are no *cooperative* or *inhibitory* interactions among parts of the system that are relevant to its behavior.

Different computational mechanisms may display different non-aggregativity characteristics. For example, a connectionist network may change only quantitatively under addition or subtraction of nodes (hence its gradual degradation); a classical von Neumann computer would suffer a dramatic change of behavior under removal of its main memory or central processing unit. Yet there might be lesser inhibitory interactions in classical computers (they exhibit carefully engineered modularity in their structure) than in connectionist nets. Note that a pail of water would fulfill W1–W4 with respect to most of its standard behaviors.[10]

5.2 Multiple Realizability and Modularity

Traditionally, multiple realizability was considered to be *the* feature of computers and, hence, of cognition in general. It was used to argue against reducibility of psychological theories to neuroscience and to vindicate their autonomy (Fodor 1974). But in recent years, such a position seems to have lost its plausibility. For one, multiple realizability might be much more limited a phenomenon in the biological worlds than classical functionalists suspected (see Bechtel and Mundale 1999). Second, it is hard to find noncontroversial examples of actual multiple realizability in neuroscience or psychology (Shapiro 2004, 2008; Polger 2008). So it will be useful to define the notion and see how it might be relevant to computation.

A functional capacity is multiply realized if, and only if, its occurrence is due to different realizing structures. For two structures to be different, the causal model of the capacity has to be sufficiently different in each of the cases. That is, it has to feature different relevant variables interrelated in different ways. For example, the color of a corkscrew is causally irrelevant to its capacity to remove corks, so differences in color do not entail differences in realization. In contrast, an electric corkscrew and a manually operating one are causally different with respect to capacity. These are two different mechanisms realizing the same capacity.

It is important to see that claims of multiple realizability depend strongly on how fine-grained the description of the capacity is. If we define a capacity of the electric corkscrew as "removing-corks-electrically," then obviously the traditional manual corkscrew will not have this capacity. Function ascriptions depend on epistemic interests, so what counts as different under one ascription may be the same under another. Even if we can formulate some general norms of functional ascription and require that they should be as general as possible (see, e.g., Millikan 2002), the level of generality itself depends on the theory in question. This has immense significance for understanding multiple realizability of computation.

So when are computational mechanisms realized differently? They need to share the same functional capacity. But there are at least three ways to understand difference in realization, depending on how we define sameness of the capacity relative to the focus of the theory in question. First, one could say that they need to share the abstract model of computation matched with the causal model and differ in some other realizing structures; that is the complete engineering blueprint. To put the point simply, the isolated level needs to be the same in these mechanisms. It is enough to replace a memory chip in a laptop with a faster one to achieve a different realization. The functional capacity of the computational mechanism includes not only a mathematical specification of the computation but also all kinds of other considerations relevant to its execution.

Such a specification of the capacity might be useful when precise timing of processing is important. For example, in cognitive psychology, response times are important when describing capacities—and this involves not only general complexity considerations (van Rooij 2008) but also empirical evidence about processing speed.

Second, we may identify the functional capacity with the mathematical function of the abstract model of computation no matter how it is implemented. This seems to be the preferred way of classical functionalism to talk of computational operations: what we care about is only input and output information. The realizations would be different only if they have different causal models associated with their fine-grained models of computation. This is what Scheutz (2001) seems to have in mind when he speaks of the divergence between causal and computational complexity. In his opinion, when two different compilers produce different machine code for a program, the program still has the same computational function but a different causal complexity of execution.[11] Such is also the theoretical interest of equivalence proofs for various machines in computability theory. Now, two instances of my laptop, before and after the replacement of the memory chip, count as the same realization; the causally relevant variables in the model of the capacity are exactly the same, and memory speeds are as irrelevant as the color of a corkscrew.

Finally, one may specify the function of the mechanism as the complete causal model on all levels of its composition (or, at least, on its isolated and constitutive level). But if computational capacities of mechanisms are specified in such a way, it becomes doubtful that there is anything multiply realized at all. Briefly, if we include the complete causal profile of a mechanism in the specification of its functional capacity, then it is no longer possible to realize it in a different way. After all, two mechanisms cannot be *causally* different while having the same complete causal profile. But if we seek replacements of computer parts, we need to take into account quite subtle considerations, such as the one that memory chips of two different brands, even if sold with same specifications of voltage, pinout, volume, and speed, might not work correctly together. So there is use for such a specification of the capacity as well.

On the first rendering, we treat only the computational specification of the mechanisms as crucial for its function. For multiple realization, we need to change the causal structure of the constitutive level without changing the fine-grained specification of the isolated level. If only the memory chips of the laptop have different speeds, they would change speed at the isolated level as well, so their replacement would not produce another realization at all. On the second rendering, multiple realization abounds but is also trivial. We know a priori that there is a plethora of different algorithms that compute the same function. On the third, it becomes virtually impossible to find different realizations of the mechanism. In other words, it is no longer clear why multiple realization is so crucial. In all cases, it seems more apt to talk of instantiating similar functional classes of mechanisms than of realizing complete mechanisms. For this reason, a slightly different notion seems to be more adequate—namely, that of *substrate neutrality* (Dennett 1995, 50).

Let me show why multiple realizability seems irrelevant to computers. Take two logically identical computers, such as IBM 709 and IBM 7090. These computers shared the same logical diagram but were built, respectively, with tubes and transistors, and they had different physical properties (this example is used by Wimsatt 2002, 203). Transistors worked faster than tubes, so the speed of operation was different. On the first and third rendering of multiple realizability, IBM 709 and IBM 7090 were two different *types* of systems with *different* system capacities; on the second, they were the *same* computational systems realized in the *same* way, as tubes and transistors do not make any difference at all to the software, which was the same all the way down (see figure 2.8). In neither case were they realizing the *same* function in a different way. Thanks to the isomorphic models of computation in both mechanisms, they were comparably similar,

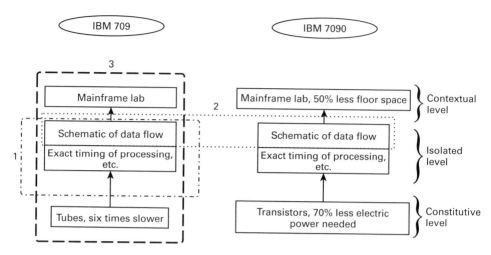

Figure 2.8
Different ways of understanding the computational capacity of IBM 709 and IBM 7090 as (1) the isolated level, (2) only the logic of the data flow, or (3) complete causal model. In case (2), only the logical level is included in the comparison of the computational capacity, so there is no difference in *realization* properties between both systems (transistors and tubes are as irrelevant as colors of corkscrews). In case (1), different speeds change the function of the isolated level, so just as in (3), where the complete causal model of IBM 709 is different from IBM 7090, we don't have multiple realization.

but not isomorphic under substitution of parts (one cannot simply replace a tube with a transistor). This kind of comparable similarity is exactly what is meant by substrate neutrality, which may involve neutrality with respect both to the constitutive differences (when the capacity to compute includes other causal factors) and to the whole causal organization (the classical multiple realization).

The proper specification of the computation at the isolated level is usually couched in substrate-neutral terms. In other words, the states play their causal roles because they are reliably differentiated from other states of the substrate; so "substrate-neutral" is not to be equated with "substrate-independent." This is why one and the same computation can be implemented, or realized, by a number of physically different mechanisms as long as they preserve the relevant information-bearing structure *and* dynamics.

As will be evident in chapter 3, the substrate neutrality of models of computation is a sine qua non of the explanatory value of the computational modeling of information processing. It is the most general criterion of explanatory applicability of computational models to cognitive processes. Naturally, nothing prevents us from describing a computational process in a non-substrate-neutral way; such a description, however, may not capture the informational regularities when a variety of substrates are employed. For standard computers, a substrate-neutral way of talking might be very natural, for even when different physical processes underlie the storage of information on optical disks and USB flash drives, they are encapsulated from the computational machinery of my laptop by an appropriate interface that presents the information simply as strings of bits, without distinguishing their sources. If the capacity of a mechanism relies on the informational regularities, we have a reason to believe that similar encapsulation will be possible. In other words, the physical differences between flash drives and optical disks notwithstanding, the behavior of both mechanisms is sufficiently similar so that we can safely abstract away from them to describe, explain, and predict a large class of information-processing capacities. At the same time, we know we cannot insert an optical disk into a thumb drive, and in some contexts these differences matter.

It is a platitude that artificial computational systems are built in a hierarchical fashion, and the notion of level is often used to describe them.[12] Lower-level computational mechanisms might be modular and separable from the higher-level infrastructure to enable better maintainability. Although biological computation might be much more complex, and the mechanisms may be much more polyfunctional than in artificial systems, there are evolutionary arguments to the effect that there will be at least some level of modularity (or near-decomposability). One of the earliest advocates of this view was Herbert Simon (1962). Yet these arguments do not undermine the obvious fact that parts of evolution-generated mechanisms are reused, and that makes them hard to decompose. A lower-level mechanism might become essential simply because it emerged early on in history, like the QWERTY keyboard, which is used in modern-day computers because it first appeared in the mechanical typewriter (for an analysis of the phenomenon of generative entrenchment, see Wimsatt 2007, chapter 7). Additionally, higher-level functional organization may not correspond directly to the elements of the lowest level of the computational hierarchy (especially if it is nonaggregative), and attempts at localizing a higher-level function on the lowest level might constitute a reductionist localization fallacy (see Bechtel and Richardson 1993 for an extended discussion). In short, the more complex the functional organization, the more interactions tie internal and external functional structure together (Wimsatt 2002, 211).

Modularity, however, is not a strict requirement for computational systems. It is just an architectural feature that we may reasonably expect in reliable mechanisms. Note also that this kind of modularity is not to be confused with what Fodor referred to by that term (Fodor 1983); it need not involve any kind of informational encapsulation or domain specificity or suchlike. (This is not to say that this architectural notion is not actually presupposed in cognitive research; cf. Miłkowski 2008; for a useful review of the notions of modularity in psychology, see Barrett and Kurzban 2006). There is a third sense of "modularity" that is relevant to computational mechanisms (Kuorikoski 2008). Namely, the interventionist theory of causation imposes a modularity constraint on causal systems; a local intervention on a part of the system should not change other causal relations in that system. This constraint has generated criticism, since many ordinary causal systems seem to break this condition (e.g., a carburetor) (Cartwright 2001, 2002, 2004). A similar worry has been voiced by antirepresentationalists who claim that representationality requires modularity (Wheeler 2005); the fact that cognitive systems are distributed and holistic is supposed to undermine their modularity.

My reply to both objections will be the same. Cartwright and Wheeler presuppose a stronger interpretation of the modularity claim than what either interventionism or the decomposition strategy require. More specifically, Cartwright complains that parameters in structural equations that describe the operation of the carburetor are not independent of each other, which is what she means by "modularity." Her point is true of most structural equations, however; this is why they are integrated models in the first place; they are not *parameter-modular*. As Kuorikoski notices, these causal *variables* are independent at the same time, but this is *variable-modularity*:

When a mechanic is fixing a combustion engine, he presumes that when replacing a damaged part with a part that is functionally equivalent in the relevant ways, the replacement should not affect the causal properties and functions of the remaining parts. However, notice that this ontological locality requirement is different from the modularity requirement for causal models, according to which the functional forms or parameter values of the separate dependencies in the model should not change when values of some variables are changed by an intervention. The possibility of intervening on the values of *variables* imposes a weaker modularity requirement than that which would be required for the possibility of independently changing the functional forms or parameter sets that represent individual submechanisms. (Kuorikoski 2008)

Cartwright simply assumes too strong a reading of the modularity requirement in her criticism of interventionism. There are causal structures that are more tightly integrated and where a change of parameters always induces changes in other parameters; but there are still interventions on variables that preserve the independence requirement. Notice that the difference between a variable and a parameter is itself context-dependent:

The distinction between variables and parameters is usually well defined in the context of a given model. It is largely a matter of the aims of modeling, a consequence of the somewhat arbitrary but unavoidable division between what is endogenous and what is left exogenous and what is thought of as changeable and what fixed. Causal relations are change relating and what are perceived to be the relevant causal relations depends, to some extent, on what kinds of changes are thought to be relevant. If the modeler is interested in how some features of a system are determined by other features of a system, she endogenizes these features as dependent variables. If some features are taken for granted or possible changes in these features are thought to be unlikely or irrelevant, they are left as parameters. (Kuorikoski 2008)

In short, mechanism does not require extreme modularity and independence of component parts. Similarly, mechanistic decomposability does not entail that there is no integration or that boundaries of the system are completely impermeable. It is just that information-processing systems need to have multiple information-bearing states that change independently of each other; otherwise, they carry no information. In this sense, they are complex. The causal structure of the flow of information is likewise complex: if information flow is not trivial, it will involve multiple independent causal variables. Again, the degree of decomposability may vary, but it is reasonable to expect that reliable systems will be near-decomposable. If they are completely nondecomposable then, indeed, they are not mechanisms (Craver and Bechtel 2006). But there is no reason to think that mechanisms cannot be distributed or holistic. They just cannot be so extremely holistic that they would have to be modeled with a single causal parameter; neither can they be extremely distributed so that there would be no integration whatsoever. In both extreme cases, the physical structures are not complex integrated systems. As cognitive systems are always integrated and never completely disintegrated, there is no clash between the extended cognition theories and mechanistic explanation.

Early cognitive science often supposed that, because of the hierarchical structure of computational systems, the neuronal level could be left underspecified; consequently, most research was done on the top level. Yet even when the lower-level systems exhibit substrate neutrality, knowledge about the way they work may restrict the search space of possible computational models that can be used to explain biological cognition, as some models might turn out to be biologically implausible. (Shapiro [2004] insists that we may also predict psychological features from neuroscientific evidence.) In current computational neuroscience, however, computational systems are also analyzed as a hierarchy of mechanisms. For example, Rolls describes the brain as a large-scale computational system that has to be understood on multiple levels: (1) on a global level, where many neurons operate to perform a useful function; (2) on a molecular level that gives insight into how the large-scale computation is implemented in the brain; and (3) on the level of single neurons, to show what computations are performed by the computational units of the system (Rolls 2007, 3). What is considered to be the bottom level of any mechanism varies according to the focus of a theory (Machamer, Darden, and Craver 2000), and this is true also of computational mechanisms. Even if early theories left implementation of cognitive mechanisms underspecified, what they took to be the bottom level did not need to be identified immediately with any neural structure (e.g., one of the levels that Rolls mentions). For example, if a psychologist has a reason to suppose that deliberate problem solving is implemented by robust regularities in conscious thought, he or she may posit these regularities as the constitutive level of the computational mechanism. Indeed, the founders of cognitive psychology believed that "the sequence of operations executed by an organism and by a properly programmed computer is quite different from comparing computers with brains, or electrical relays with synapses, etc." (Miller, Galanter, and Pribram 1967, 16, n. 12), and they focused on the first one (ibid.; Simon and Newell 1956; Simon 1993).

I am not implying that the approach of early theorists is inadmissible from the mechanist point of view, and no kind of neural reductionism is required by my account of implementation. At the same time, it is quite obvious that today, the cogency of high-level symbolic regularities in cognition that are not in any way theoretically unified with any lower levels of the cognitive system organization is rather disputed. For natural cognitive systems, integration with biological theories is desirable; we want to know how it is *biologically* possible for cognitive apparatus to display symbolic regularities, as many people seem to deny that they actually do exhibit such abilities (e.g., Piaget's symbolic explanation of behaviors can be overridden with a motor one [see Thelen et al. 2001], so we cannot posit a stable symbolic level as a bottom one without further justification). Similarly, artificial cognitive systems need to be unified with materials engineering to understand the operation of robots and such. It simply seems that there are low-level constraints on symbolic theories that are important for understanding the structure of symbolic information processing, and they are found only by integrating several fields of research. But this is just a heuristic that is generally thought to be useful today; it is not to be confused with "conceptual truths" against symbolic-level theories.

The notion of a level of organization of a mechanism should not be confused with other kinds of levels that are introduced in computer science. In particular, the level of description might not correspond to any organizational level at all (Floridi [2008b] calls such levels "levels of explanation," but their purpose is not necessarily explanatory). For example, an algorithm that is actually implemented using the machine code of the CPU in my laptop might be described on several various levels of abstraction using different descriptive frameworks. Some of these descriptions might even be parts of the history of the machine code: namely, the source code in a high-level language is almost never executed as such on a computer. (I added "almost" because, in principle, you could create a machine that runs a high-level language directly.) It is first translated into machine code with the use of interpreters (during the run time) or compilers (off-line). The compiler may further optimize the resultant machine code so that the reverse operation is difficult—if not impossible—to perform without the knowledge of what optimizations were applied to the source code. For this reason, the high-level source code of software is not really a direct causal factor in a computer, even if, for practical purposes, when precision is not required, we might say that a computer executes a program written in C++ or Java. It is only indirectly executed via compilation or interpretation. What is more important here is that some of the high-level structures may have no counterparts in the machine code. For instance, a variable in Pascal code might not figure in the machine code (Scheutz 2000); a loop might be translated into a series of instructions without any conditions being checked—for example, to speed up some critical operations. In particular, there is no relation of logical supervenience between high-level code and machine code. Machine code may vary without there being any changes in the high-level code, and the high-level code may vary without there being any changes in the machine code (for definition of logical supervenience, see, e.g., Kim 1993 or Chalmers 1996b). The supervenience basis would have to include a complete specification of how the interpretation process works. This stands in stark contrast to mechanistic levels; high-level entities are not just abstract, less detailed descriptions of lower-level entities or structures, and there is no process of interpretation that needs to be included.

6　Possible Objections

Classical antirealist objections to computational descriptions were voiced by John Searle (1992) and Hilary Putnam (1991), who argued that there is no way to distinguish computational processes from noncomputational ones, as there is no effective way to individuate them. On the mechanistic account, the individuation seems to be vindicated, and the computational theory of the mind in its mechanistic guise (as I showed in section 5) is in no danger of being vacuous or trivial. But there are some further objections that need to be answered.

6.1　Syntax Is Not Physical

The argument formulated by John Searle (1984), which became a basis for many subsequent criticisms, runs as follows (I use the rendering of the argument from Chalmers 1996b):

1. A computer program is syntactic.
2. Syntax is not sufficient for semantics.
3. Minds have semantics.
4. Therefore, implementing a program is insufficient for a mind. (Chalmers 1996b, 326)

Obviously, in my account, this argument is a non sequitur: computer programs could be syntactic, but I do not presuppose either that programs are necessary for implementation, or that an implementation of a program is the same as the program itself. As Chalmers has noted, Searle commits a simple fallacy: he confuses the thing which is realized (an abstract specification) with its realization (basically, a physical process). It is as if cookies could not be crispy as no recipe (as abstractly conceived content, not as the physical bearer of the recipe) itself is crispy!

A deeper interpretation of Searle's claim would be that no physical description would suffice to establish that there is a mathematically described regularity in the world. In other words, computational descriptions are never descriptions of natural kinds; they are purely verbal. Yet the arguments he uses regarding this claim seem to undermine the uses of mathematics in physics, including the act of measurement (McDermott 2001, 171; Dresner 2010; Miłkowski 2012). Were the argument correct, all mathematical properties containing syntactic and formal properties (and if there are any formal or syntactic properties at all, complex mathematical structures ascribed by equations in physics are among them) would be outside the scope of physics, as syntax is not intrinsic to physics (or so Searle claims). But that would mean there is no justification for quantitative ascriptions to physical processes, and we would have to reject all our best scientific theories. I do not know of a better *reductio* of this sort of armchair theorizing.

I also do not presuppose that semantic representation is necessary for computation, and I do not claim that computation suffices for explaining the whole range of semantic properties. In other words, this argument, were it valid, would still be beside the point.

6.2   Morphological Computation

In embodied cognitive science, there is a developing new field of "morphological computation" (Paul 2006; Pfeifer and Iida 2005; Pfeifer, Iida, and Gomez 2006). There is no clear-cut definition of the term, and it seems that it is used in two ways. The first stems from the observation that morphology of robots sometimes plays a role that would otherwise be computational: "It turns out that materials, for example, can take over some of the processes normally attributed to control, a phenomenon that is called 'morphological computation'" (Pfeifer and Iida 2005). So, for example, the pressure-difference receiver in crickets (Webb 2008) introduced in chapter 1 would be playing a morphological-computational role on this account, as it obviates the need for the filtering that would otherwise be required for the sound of the male cricket to be processed by the female cricket. Now, doesn't my account render such devices uncomputational? Or if it does not, doesn't this make just any mechanism computational?

Do the ears of the cricket process information? Just like any sensor, they do; as Pylyshyn (1984) would say, they are not merely transducers of energy into information. The information is already out there, at least in terms of the structural-information-content (MacKay 1969). The sound waves have a limited number of degrees of freedom, and it is quite clear that they encode information without there being any sensors. If there is a receiver, we know that there is selective-information content as well. In this case, the receiver is the pressure-difference receiver with four eardrums. So there is no doubt that we have information processing here. There is a mechanism with an evolutionarily designed organization of functional parts that were selected to process the auditory information. The biological systems and subsystems have clear boundaries (cell membranes, etc.) and can also be clearly individuated. The function of information processing can be described as the filtering of sound. But, admittedly, this happens thanks to the morphological features of the sensor that appropriately encode the signal in a physical, analog way. There isn't any in-principle difference, however, between this use of physical features of morphology and an engineer's intelligent use of materials to build computer storage (e.g., the use of lasers to write and read CDs).

Morphological computation seems to be unproblematic under my account of computation, and there are more examples that show this. The morphology of the robot can be used to perform real computation—for example, a logical function, such as XOR (exclusive disjunction), can be easily implemented using robot manipulators instead of its electronic control structures (Paul 2006; Casacuberta, Ayala, and Vallverdu 2010). When a robot is designed in such a way as to use its own morphological features to compute some functions, there is absolutely no problem with treating those features as parts of the computational mechanism.

But the worry that all mechanisms may be thought to be information processors needs to be addressed. To this I now turn.

6.3 Pancomputationalism and Mechanisms

Pancomputationalism (see also endnote 5) asserts that the universe might be a huge computer computing its own future state. Although I believe it is not at all clear that pancomputationalism renders the computational theory of the mind vacuous (Miłkowski 2007)—I went to lengths to show that pails of water and planetary systems are not computers because they are systems devoid of appropriate functionality (see section 5.1 of this chapter)—a weaker version of pancomputationalism might be implied by my account. Namely, it might seem that all mechanisms are information processors. Also, a proponent of pancomputationalism might point out that there is lively research in physics into physical limits of computation and the information capacity of the universe, where computation is understood more broadly than it is in cognitive science, so my attempts to distinguish the computational from the noncomputational are misguided.

I will address these concerns in turn. First, notice that any causal process can be analyzed in terms of structural-information-content (Collier 1999), so it does not seem so far-fetched to say that all causally organized, functional mechanisms might be processing information. However, notice also that mechanisms are always specified as related to some theory as mechanisms *of* something: they are posited as entities having a specific capacity. For example, the vacuum cleaner is a mechanism for cleaning, usually, the floor and furniture. Even if you could redescribe it as a device for, say, randomizing input information encoded in the arrangement of dust particles (or something similar) or encoding input voltage used by the engine into the value of its speed, a large number of features of the vacuum cleaner would remain unmentioned. In other words, even in my account, some kind of trivial ascription seems logically possible and admissible if the target of the ascription is a functional mechanism. A traditional way to discard such candidates for computation was to say that they do not use representations; but computation does not require representation in the sense of referring to distal objects, and some trivial forms of representation (selective-information-content or structural-information-content) will be easily found in just about every physical system, so appeal to representation will not help with such cases either. My answer is that nonstandard ascriptions of computational function to all mechanisms do not bring any predictive and explanatory power. This, in turn, means that these trivial cases of computation will be subjected to an epistemological Occam's razor in chapter 3, which focuses on the explanatory aspects of computation.

What about physical systems that are not self-contained functional mechanisms? For example, physicists estimate the ultimate computational capacity of the universe or ultimate physical limits of computation (Lloyd

2000, 2002)—they regard the universe as a huge computer. So maybe pancomputationalism is not such an outrageous idea after all? Actually, it seems that it is not a part of standard theories of physics; the mentioned claims are fairly conditional. They seem to be based on the following argument scheme: if the universe were a computer, it would have computational capacity, and, considering the limits of the universe, there would have to be such and such limits to this capacity. Obviously, this conditional is true even if the antecedent is false, so the assumption of computational nature of the universe might as well be counterfactual. My account of implementation of computation could then be applied to see whether there is sufficient organization related to computing to call the whole universe a computational mechanism. The requirement that computational mechanisms be functional would be incompatible with the literal treatment of planetary systems or universes as computers: they seem to lack any design that is a product of some selection (though, in theoretical physics, one could have a theory of evolution of universes or something similar that would justify some kind of cosmological natural selection; see Smolin 1997 for a defense of such a view). I would reserve another term for such cases, namely *nonfunctional information processing*, as it seems much less confusing. All in all, I doubt that there is much explanatory and predictive gain to be had from pancomputational theories, but if it turns out that physics has a good use for them, a philosopher should not legislate that pancomputationalism—an empirical claim—is a priori false. I just think that the price for stipulating computation on the fundamental physical level should be higher; it is not a matter of only a simple mapping between a physical system and a model of computation. Again, to see whether we can go a long way with pancomputational theories, we need to focus on the epistemological aspects of computers as explanatory means, and this is taken care of in the next chapter.

6.4 Dynamical Systems and Control Theory Is Better

A final objection, voiced recently by Chris Eliasmith (2010), is that computation is not a good quantitative formalism to describe the brain. As I noted in chapter 1, his own theory is computational according to my criteria, so this claim might come as a surprise; however, what he means by that is standard digital computation. The argument is that the language of dynamical systems, especially the language of modern control theory, is better suited for describing the changes occurring in the brain over time. Now, this is not exactly an objection to the mechanistic account, which is pluralistic by its very nature; we need descriptions of mechanisms on multiple levels and from multiple perspectives. The time evolution of a mechanism is of obvious importance to its description, manipulation, and decomposition. Temporal dynamics of neurons are completely different from clock-governed sequential processing in standard digital computers. However, sequential character or clocks are not essential for positing computations in general; rather, they are related to some models of computation. For example, Turing machines are inadequate as a model of the brain exactly for this reason, as neural oscillatory dynamics are of a dramatically different nature. Here I agree with Eliasmith and other advocates of the dynamicism (Bickhard and Richie 1983, 90; Bickhard and Terveen 1995; van Gelder 1995). What I disagree with is Eliasmith's reliance on the principle that all the levels of a system should be unified by one kind of description. This is not realistic for multilevel mechanisms in biology and neuroscience: we can describe the Krebs cycle in terms of chemical processes and in terms of metabolism, and these are different descriptions. Pluralism is a feature of mechanistic explanation, and reduction to the bottom level is not required. What is required is interlevel integration that uses evidence from various levels to understand their relationships. So the brain might be described at the same time as both a processor of information and a control system. There is no deep theoretical problem in doing so.

7 Checklist for Computational Processes

Before I summarize what has been going on in this chapter, it will be helpful to make a checklist of criteria that help decide whether a given process is best described as computational:

(1) Are the initial and final states of the process linked causally?

This is a requirement that any causal process will fulfill.

(2) Are there any stages in the processing? Do they correspond to the stages of the system's activity?

If there is a complex task, then it is realized in separate steps that need to be linked causally as well, and the stages should be treated as nested computational processes in the overall process. For example, a sequence required to solve cryptarithmetic tasks won't be present in the digestive process, as the parts of the verbal protocol would not be mapped to required processes (to the best of my biological knowledge).

(3) Is the system relatively isolated?

Relative isolation is the first mechanistic item on the checklist: mechanisms need to be relatively isolated from the environment, and the boundaries of the system as described computationally should roughly coincide with those discernible on the constitutive level of the mechanism. A part of the wall hypothesized to execute Wordstar, but which has no obvious boundaries at the constitutive level that would overlap with the boundaries of the computation, is a nonstarter. A pail of water is in a better position, as there is no computation posited beforehand, so one could simply redescribe the constitutive level as a form of computation in the pail and have the same boundaries this way.

(4) Does the system support interventionist causation?

Is the system sensitive to the interventions in the way that its theoretical model requires? Are counterfactuals supported? Spurious computations fail here; a simple change of the input stream of information will need to be reflected in the appropriate changes of the whole process.

(5) Is the system organized?

Is the system a nonaggregative system? (This is a heuristic! Caveat emptor.) A pail of water and a heap of sand are not. You need to find parts that are the result of a selection process in the organization of the mechanism (e.g., an expression of genes) to ascribe real function to those parts. For example, an evolutionary process itself will lack such parts, as the very process does not have any selection mechanisms, as far as we know, and is not a cohesive structure (i.e., it has no boundaries). Nonfunctional systems are not computational. They might merely be parts of nonfunctional information processing.

(6) Is the constitutive level of the mechanism bottomed out?

The constitutive level of the computational mechanism needs to have parts that are not identified in the computational way only. A merely functional cheater-detection module that has no neural base resulting from the brain being decomposed into parts (e.g., thanks to natural interventions) is discarded here.

(7) Does the trajectory of causal states correspond to the computation model?

The whole trajectory of states of the mechanism should correspond to the model of the computation in a fine-grained fashion. All significant changes in the mechanism parts that are theoretically significant should be reflected in the computation as well.

(8) Is the computation model complete?

Flow diagrams are nice, but they do not execute as programs; they might sketch a mechanism, but a complete description requires a full specification of the computational model. If the system to be described is programmable, the specification should be couched at least in terms of the pseudocode that is run by the system. If the code is complex, it might be described on some level of abstraction to remain informative, yet there should be technological ways to use the real code as well.

8 Summary

In this chapter, I reconstructed several accounts of physical implementation of computation. The first one, the simple-mapping account, is prevalent in early functionalism, yet it remains too broad and is vulnerable to objections articulated by Searle and Putnam. Alas, semantic or formal symbol accounts are not huge improvements over simple-mapping accounts. Framing computation in terms of information processing can save some of their insights; however, this is not completely satisfactory, as individuation of entities that process information is still lacking. The more robust account, the mechanistic one, supplies individuation criteria in terms of constitutive relevance and is safe from danger of trivialization. It is much narrower. As it will turn out in the next chapter, it also has better explanatory properties.

3   Computational Explanation

This chapter explores various explanatory uses of computational modeling in cognitive science and investigates the norms for assessing whether a given computational explanation is effective or merely constitutes a spurious redescription of a phenomenon. I will review three theories of explanation that might elucidate computational explanation, namely (1) the covering-law theory, (2) functionalism (including Marr's three-level account), and (3) mechanism. I conclude that the mechanistic framework is best suited to the task of elucidating computational explanation. However, this is not to say that I argue for a complete denial of other explanatory frameworks; on the contrary, mechanistic explanation might be fruitfully construed as a natural extension of the covering-law account. I also see a role for explanatory unification of various models.

1   Methodological Advantages and Uses of Computational Models

Surprising as it may seem, one can still find people who believe that mathematical modeling in psychology has failed. This is so in spite of the many well-known advantages that computer modeling has when used in science. It might be a good idea, then, to remind ourselves, if only briefly, why this opinion is unjust. The most obvious gain that comes from employing mathematical modeling in any field is associated with the fact that it promotes clarity and precision. While it is true that theories expressed solely in terms of natural language are often more accessible than formal models, especially if one lacks a background in mathematics, such theories remain notoriously difficult to test; words are malleable, and it is not always clear how to interpret them. Quantitative theories, on the other hand, yield much more accurate predictions, which, given the availability of proper measurements, lend themselves to empirical evaluation. Failing that (if, say, the theory does not supply any metrics), they are still more rigorous than qualitative accounts, which means that they lead to fewer verbal discussions.

Building strict mathematical models also helps scientists to overcome some inherent limitations of human thinking (Farrell and Lewandowsky 2010). For example, to properly understand and evaluate a verbal claim about parallel distributed processing, a researcher would have to rely heavily on his working memory; given working memory's limited capacity and the high complexity of such problems, the process of evaluation would be extremely time- and energy-consuming, while its result would likely be wrong. Such difficulties are easily sidestepped thanks to computer simulations. Inconsistent analogical reasoning, typical of informal thinking, is not a problem for computer models, whose behavior is the same every time. Creating many competing models of a phenomenon, a method common in cognitive psychology, reduces confirmatory bias (i.e., the tendency to seek evidence that confirms, rather than disconfirms, a hypothesis). At the same time, the fact that computer code can be shared easily facilitates fast and reliable communication among researchers.

This is not to say that one cannot be sufficiently exact in natural language. If that were true, I would not be writing these sentences in English. What I mean is that, unlike linguistic representations, mathematical models cannot be vague or ambiguous, and implementing them imposes further constraints on clarity. For example, a verbally articulated claim that one cannot master the past tense of verbs without grasping a set of explicit rules is admittedly harder to assess than a computer model if one does not define what "grasping" means (recall the discussion over the distinction between following a rule and behaving in accordance with a rule, which were sparked by later writings of Wittgenstein; cf. Kripke 1982). The mere existence of a model that can learn without appeal to a set of rules (Rumelhart and McClelland 1986) can therefore go a long way toward disproving the verbal claim. In other words, existence proofs, or demonstrations that a given task can be accomplished in a particular way, have a direct application in overthrowing impossibility claims, even if they never prove that human cognition works along the lines of the model in question.

Before I go on, some terminological clarifications are in order. I assume that scientific theories are general representations of how the world is.

They can be couched in a variety of formats: as sentences in natural language, sets of equations, statements in a logical calculus, diagrams, animations, computer programs, and so forth. Because of their format, representations vary in their accessibility to humans (for classical results concerning the difference between diagrams and words, see Larkin and Simon 1987), which means that equivalent information might be easier or harder to understand depending on whether it is presented verbally, on a diagram, or as a computer simulation. Since the cognitive capacities of scientists and lay people alike are limited, we should use the representations that are easiest to understand or produce. They should also be sufficiently clear and rigorous. It is difficult to fulfill these requirements in conjunction, and some trade-offs might be inevitable; we should make our choices in light of the scientific goals we are pursuing. Sometimes a rough diagram or a linguistic metaphor may be enough for our purposes; at other times we need the complete mathematical derivation of a proof.

A model in cognitive science differs from a scientific theory in that it is not cast in natural language, and its scope need not be general. I do not want to go any deeper here, as we already have in the literature a number of distinctions pertaining to models in science (for a terse review, see, e.g., Webb 2001). Although theories in psychology and cognitive science tend to be relatively narrow in scope, I do not see much risk of confusing them with models. Generality in this case should not be equated with the scope of phenomena that a theory describes, explains, or predicts; it also resides in the generality of the principles that drive model building, such as the background assumption that there is an important distinction between short-term and long-term memory. One could provide more precise notions of model and theory, but it would be impossible to do so without appealing to a particular theory of explanation. Because the use I am making of these concepts is intended to be fairly neutral as to the broader explanatory framework in which they may be embedded, I deliberately keep them unspecific. What I deny, however, is the central tenet of the so-called received view on the structure of science: namely, that a theory has to be a set of propositions in a formal calculus. Nevertheless, it can be presented in such a form.

Simon and Newell (1958) went as far as to predict that all psychological theories would be presented as computer programs; in my terminology, their prediction would be equivalent to the claim that theories would enable researchers to create computational models presented as computer programs. Newell and Simon certainly wanted psychology to be so. However, even they, despite being regarded as radicals by mainstream psychologists (Baars 1986), did not restrict psychological theorizing to actual program writing. More specifically, even if we leave only "computational models" instead of "computational models presented as programs" in my rendition of their claim (given the existence of connectionist simulations, which are not programs in any straightforward sense), adopting this kind of constraint would exclude many bona fide explanations, such as an account of certain general features of hypnosis appealing to complexity considerations of information processing (Dienes and Perner 2007). We do not need complete programs to assess complexity. It is enough to know the complexity of the algorithm; the requirement to write programs would be more exacting than the standards applied to complexity theory in computer science. As Boden (2008, 742) stresses, the fact that Dienes and Perner's "theory of mental processing has not been expressed in the form of a computer model does not mean that it is empirically ungrounded or untestable." This is why Simon and Newell added that theories could also take the form of "qualitative statements about the characteristics of computer programs" (Simon and Newell 1958, 8).

This qualification is important. Early cognitive science pursued several explanatory strategies. One was cognitive simulation, which proceeded via building complete models, and another was constructing artificial neural devices such as the Perceptron (Rosenblatt 1958) or robotic animals (Walter 1950). Still another was what was to become cognitive psychology, called information-processing psychology at the time (for exemplars, see Miller 1956; Broadbent 1958; Sperling 1960). Information-processing psychology did not build complete implementations, but investigated the structure of psychological processes as based on information-processing considerations. By using methodologically strict experimental evidence, it paved the way for other kinds of cognitive research (see Baars 1986).

To be sure, complete models are not without advantages. An ability to create the computer simulation of a phenomenon shows that the phenomenon is understood fairly well: the description is precise and does not contain any gaps. This is a virtue of both cognitive and neuroscientific computational modeling:

A test of whether one's understanding is correct is to simulate the processing on a computer, and to show whether the simulation can perform the tasks of memory systems in the brain, and whether the simulation has similar properties to the real brain. The approach of neural computation leads to a precise definition of how the computation is performed, and to precise and quantitative tests of the theories produced. (Rolls 2007, 1)

If the simulation works, it also supports the supposition that the theory is consistent; inconsistent theories yield faulty models, and from theories that are both incomplete and inconsistent one cannot (usually) derive any models at all (Baars 1986, 181). Moreover, although Pinker and Prince (1988) criticize Rumelhart and McClelland's model of past-tense acquisition quite harshly, they admit that its value lies in its being implemented. It is the implementation that enables researchers to assess the properties of the model that the original investigators had not initially contemplated. For example, people respond faster to a past-tense form of regularly pronounced present forms than to inconsistently or irregularly pronounced ones. This property could not be tested in classical or hybrid models because they were unimplemented (Shultz 2003, 97).

With complete working models, we can tweak their parameters and see what predictions we get; Cleeremans and French (1996) call this process the "probing and prediction" of models. Interactive modification of the model enables researchers to see the exact changes that their theory implies. This, in turn, can lead to an improvement on a previously unsatisfactory model. In other words, modeling is a cyclical process that starts with a rough prototype that is gradually enhanced to deal with the data and is eventually tested on data that was not used to build the model. As Lewandowsky observes, "simulation of a verbal model can reveal previously hidden insufficiencies, and . . . simulations allow for experimentation and modification until known empirical benchmarks are accommodated" (1993, 237). This, however, does not mean that models are fitted to data and confirmed automatically; the benchmarks and evaluation methodologies should be as strict and precise as the models themselves. The interactive use of models in their development cycle justifies the claim that implemented models are generally more valuable than unimplemented ones (Mareschal and Thomas 2007).

Before exploring the explanatory uses and norms of computational explanation, it is important to note that all models as such are idealizations. Most models are computer simulations; that is, they are supposed to represent the phenomena. All simulations have finite precision and can be used to predict only some aspects of the modeled phenomenon. Whether the simulation is just a representation of the phenomenon on a computer or is supposed to instantiate the same kind of information processing as the real phenomenon, it will be selective. It might be tempting to say that models are abstractions, but this is not enough: they are specially crafted to account for a special range of features of the phenomenon in question. As Leszek Nowak always stressed, this selectivity is characteristic of idealization, whereas abstraction is simply the discarding of some properties that may be essential or not.

There are also several kinds of idealization procedures that can be applied when modeling a phenomenon. *Galilean idealization*, for example, consists in distorting theories so as to simplify them; *minimalist idealization* involves representing only the core causal factors relevant to a phenomenon; *multiple-models idealization* (MMI) is the building of multiple, mutually incompatible models that only collectively satisfy our cognitive needs (Weisberg 2007). Arguably, all three kinds are found in computational cognitive science, with Galilean and minimalist idealizations being the most common. One important point of employing multiple models, as I mentioned earlier in this section, is to prevent confirmatory bias.

Moreover, rather than explaining raw data, scientific theories, including those that rely on computational modeling, are typically tested against what Suppes (1962) calls *models of data*, which present experimental results in a statistically corrected and regimented manner often informed by a host of nontrivial theoretical assumptions. Chomsky's (1959) critique of Skinner's theory of language provides an especially clear illustration of just how important models of data might turn out to be. According to Chomsky, Skinner's construal of verbal behavior does not accommodate a fundamental feature of natural language—its productivity. (The set of grammatically correct English sentences that can be produced or understood by an idealized native speaker is infinite.) This is a valid objection even though, strictly speaking, productivity cannot follow from any set of empirical data gathered by human researchers.

Let me relate idealization to computational descriptions. As Drescher (1991) rightly observes, no physical computer ever has unlimited memory. But, even so, we have reason to describe some physical systems, in an idealizing way, as computationally equivalent to the universal Turing machine even though, physically, they cannot be anything but finite-state automata. Drescher writes: "There are no precise rules governing the suitability of this idealization; roughly, the idealization is appropriate when a finite-state automaton has a large array of state elements that it uses more or less uniformly—elements that thereby serve as general memory" (1991, 38). In other words, we assume counterfactually that the memory of the machine is unbounded, and for this reason it is not just an abstract, general description, but a counterfactual distortion.

The same may apply to all the other models of computation as related to physical machines. For example, a proponent of hypercomputation, such as Siegelmann (1994), might argue for describing a neural network as a system performing computations over genuine reals, despite knowing full well that all physical systems are limited and perfect precision of measurement is, strictly speaking, physically impossible. She might still want to idealize the system in a Galilean fashion. (In other words, there might be epistemic reasons for going beyond raw data; the epistemic considerations are therefore quite different from ontological questions of implementation that I dealt with in chapter 2.)

To evaluate various computer models, we need a theory of idealization that would allow us to assess the epistemic value of simulations. Nowak's approach to idealization (Nowak 1980; Nowak and Nowakowa 2000) stresses the counterfactual nature of idealizational explanation: instead of specifying all the factors that influence the explanandum phenomenon, it focuses on the essential factors, which is true of Galilean and minimalist idealization, but not MMI.

How to decide which factors are essential? In Nowak's view, the notion of essentiality remains primitive; nonetheless, Paprzycka (2005) offers two explications of it. Roughly speaking, they correspond to the depth and breadth of a simulation that Newell and Simon (1972) were concerned with. Depth and breadth are roughly equivalent to common measures of information retrieval (i.e., precision and recall). As Newell and Simon remarked, there must be a trade-off between the two: it is trivial to redescribe the observational data and get 100 percent precision this way, but with a loss of generality (this is also called overfitting); it is also quite easy to make the simulation so general as to match any observational data whatsoever, which would leave the model devoid of empirical content.

The explanatory methodologies explored below will help to fill in the blanks in the general scheme of evaluation. I will argue that the most robust norms of computational explanation are supplied by the mechanistic account of explanation. At the same time, I want to do justice to explanatory techniques that do not square well with mechanistic methodology. Although, for example, some Bayesian models appear to be incomplete sketches of explanations, they might be regarded as fruitful idealizations in certain theoretical contexts. We therefore need to explore a number of accounts of explanation to see what requirements they articulate. Yet before I turn to this task, it is worth showing that computational models are tested to ensure their empirical adequacy.

2   Testing the Models

Let me start out with what seems to be the most general condition any computational explanation should meet. Namely, if the process is best understood as transforming information, then the computational model may capture its essential characteristics. Computational models will therefore be applicable to such processes, and they will be interpretable not only as useful tools to calculate the evolution of a process in time, but also as literal descriptions of the essentially similar process of information manipulation that has the same structure. In the case of a weather simulation, for example, we do not suppose that clouds process information about the air pressure; similarly, planets do not calculate the equations that govern their movements. But in the case of a computer simulation of addition of integer numbers by my calculator, the calculator itself computes the same (or a sufficiently similar) function. In computational psychology and cognitive science, this kind of modeling is of utmost importance. Computational models, in other words, capture the information-processing characteristics of cognition.

How can one tell whether a physical process deals essentially with information? Here an intuitive idea of substrate-neutrality might come in handy as a kind of informal test. In chapter 2, I introduced the example of IBM 709 and IBM 7090. Although logically identical, the two computers were built differently: one consisted of transistors and the other of tubes. Despite this difference, there is a clear epistemic gain in describing them as logically the same if one's explanatory goal is to understand how the computer solves information-processing tasks. If, on the other hand, one needed to explain the power drain, the logical identity of function would be irrelevant. In the case of cognitive explanations, as I argued in chapter 1, we care about information, and this is why the function of information processing is essential. Information processing as such is substrate neutral: it suffices to find a substrate that can replicate an equivalent number of distinctions (structural-information-content in MacKay's [1969] terminology). Obviously, substrate neutrality constitutes merely a necessary condition for computational description having explanatory value. Accordingly, although money is substrate neutral, its purpose is not to process information, but to help economic exchange—hence it is not an information processor.

The intuitive test of the applicability of computational explanation is to ask whether the process under consideration might be realized *the same way* in a different information-processing medium. If not, then the explanation is spurious. This is a basic condition for the meaningfulness of any computational modeling of cognition that is employed to capture the essential causal structure of the process rather than serve merely as a tool (as in weather or earthquake simulations). Note that multiple different realizations of the capacity are quite irrelevant, and, as I insisted in chapter 2, multiple realizability should not be confused with substrate neutrality.

The mathematical rigor of computational explanations would be lost if they were to be tested only informally. For example, one disadvantage of the Turing test is that it relies on the subjective assessment of the judges; there are no objectively or intersubjectively reliable metrics that can determine which programs get closer and closer to passing the test and which do not. Note that the original Turing test does not require that judges agree, and leaves the possibility open that they simply toss a coin to decide; this subjective bias is inevitable. Even if Turing rejects as absurd the idea of performing a survey to discover whether machines can think, in the end his proposal amounts to just that: using statistical methods to average out the subjective opinions of a group of interrogators. It is only the question that is replaced with Turing's variant of the imitation game (Turing 1950, 434); the interrogator is to decide whether he or she speaks to a human being or a machine. This is why the Turing test cannot really be applied to evaluate empirical research, though, admittedly, panels of independent judges are used to assess the adequacy of computational models or to prepare training data for modeling, which is already an improvement over the original proposal.

Consider, for example, the procedure of comparing the verbal protocol with the set of steps in a production system (Newell and Simon 1972), which relies on the intuitive judgments of human experts (Fodor [1968] recommends a methodology similarly inspired by the Turing test). In such a case, creating a set of clear instructions to the judges is essential; otherwise, random differences of opinion might bias the results. In current modeling methodologies, informal validation and verification includes more than the Turing test; the opinions of experts are cross-validated, and various ways of interaction between the experts may be implemented if their judgments are in conflict (Petty 2010). Yet this is never a completely satisfactory way of validating the model. After all, if the model yields precise predictions, it would be best to test those predictions in an equally exact manner rather than depend on intuitive assessments.

More importantly, the evaluation must go beyond intuitive judgments. All human experts did for Newell and Simon was to find correspondences between the steps in the production system and the statements in a verbal protocol (a task which might, arguably, be automated using today's natural-language processing technologies to avoid random subjective bias). The overall goodness of fit, however, was not just a matter of reflecting intuitive human judgment.

The methods of empirical psychology which have grown out of the behaviorist research tradition do not square well with explanations that appeal to numerous theoretical entities (Baars 1986). This is why computer simulations were not tested empirically in the early days of cognitive science (Hunt 1968, 12–13), and there is still considerable opposition to computational models in psychology (Watkins 1990; for a reply, see Lewandowsky 1993). It is methodologically necessary to define metrics of success if computer models are to serve as explanations.

Various similarity metrics for empirical data are used to evaluate and compare models. In computational modeling, the idea that we could have many empirically equivalent but theoretically distinguishable theories is not a philosopher's fiction. One and the same phenomenon could be modeled by a classical symbolic program, a connectionist network, dynamically, or using some hybrid methods. An empirically established similarity between behavior and the model might be in the same ballpark for all of those kinds of models. So how to decide between them? An obvious way is to choose the simplest one, and for computational models the very length of the minimal model description as stated in some neutral code can serve as a standard of their complexity (for more discussion on comparing mathematical models in psychology, see Pitt, Myung, and Zhang 2002; for background on minimum description length, see Grünwald 2007). More realistic, however, is the situation where, for theoretical reasons and depending on the researchers' interests, the trade-offs between simplicity, level of detail, and scope and depth of the explanation are assessed differently. When investigating the principles of cognitive processes across several biological species, as in the theory of cognitive maps (which involves comparisons between wasps, sand scorpions, and humans), leaving out some neural details or abstracting away from them seems inevitable (e.g., Gallistel 1990, chapter 5). If, in contrast, the target of the explanation is cognitive maps *in the rat*, it would be reasonable to focus on providing a faithful, fine-grained representation of the rat's neural system—for example, by stressing the role of the hippocampus (e.g., Redish 1999; Conklin and Eliasmith 2005). As targets of theory, although arguably similar, are different, the assessment of scope and depth varies accordingly.

Worse still, theories might differ in terms of the methods they use to clean the data, so they treat different data as evidence. There are also pragmatic considerations in assessing theories—some theories are easier to use for some tasks, some for others, and so forth. For example, it is much easier to obtain data from invasive experiments on animals than on humans, so for practical reasons certain kinds of evidence will be more readily available in rat research than in human research.

Pragmatic considerations notwithstanding, we want to have tangible and relatively neutral metrics to test empirically. The first and obvious category of metrics will refer to the input and output information sequences of the computational mechanism. We can test whether the computer simulation, given some input data, emits output that corresponds to the model of empirical data. This is a usual kind of empirical test in various branches of artificial intelligence (AI)—for example, in computational linguistics. Depending on how structured the input and output data are, we can define various metrics. If the model's task is categorization, one can apply the measures used in information retrieval—that is, precision, recall, and accuracy (Baeza-Yates and Ribeiro-Neto 1999, chapter 3). These resemble the measures employed by Newell and Simon (they used "depth" instead of "precision," and "breadth" instead of "recall"). For data whose output structure is not simply a list of classifications, more complex correspondences with the model of data that encodes human performance are needed; in the field of machine translation, for example, complex similarity measures are necessary to account for possible paraphrases of the translation produced by a machine.

The empirical testing of input/output relations is therefore relatively straightforward, although defining similarity metrics for complex structures in the data might require considerable ingenuity. Input/output testing may contribute to an understanding of cognitive models as black boxes, which is rather akin to what stimulus-response (S-R) psychology had envisioned, even if, here, the input and output are not restricted to the narrow set of features palatable to the behaviorist (e.g., the physical properties of the stimulus). This is especially true of early modeling efforts inspired by cybernetics (which avoided postulating internal states of the system). For instance, the Test-Operate-Test-Exit (TOTE) model focuses on observable behavior (i.e., the exploration of the state of the environment); the "O" phase is purely theoretical and without any empirically testable consequences (Miller, Galanter, and Pribram 1967). Fodor (1968) describes such models, tested only on the level of input/output, as "weakly equivalent" to the systems they represent.

There are infinitely many mathematical models that would satisfy the loose condition of having the same finite input/output streams; it is enough to add idle steps in the computation to obtain a new model. Importantly, more significant differences in the postulated interim states cannot be decided by empirical investigation of the physical processes involved. No matter how many experiments we perform on inputs and outputs alone, they will not tell us how to distinguish between different machines with different internal states (for a proof, see Moore 1956). Obviously, one can appeal to the principle of theoretical parsimony to select the simplest models capable of mirroring the input/output relations of the system under investigation. To do this, one can use mathematical metrics, such as Kolmogorov complexity (Chaitin 1990) or minimum description length (Grünwald 2007). However, systems built by biological evolutionary processes are hardly ever optimal in this regard, so maximum simplicity of the model cannot guarantee its empirical soundness.

The upshot is that if experiments are only conducted on the level of input and output, empirically unresolvable conflicts might ensue between competing computational models of the same cognitive phenomena. The models do not have to be notational variants of each other; yet there will be no direct way to find out which one is the more adequate.

The input and output streams, however, are not usually treated as finite sequences but as exemplars of a general rule that governs their relationship. Hypotheses are then formed about the internal structure of a minimal computational model sufficient to create any instances of this relationship. For example, if the model is to take any sentence of English as its input and generate that sentence's grammatical parse as its output, it is clear that a finite lookup table mapping all possible sentences to their grammatical parses will not do. The computer model has to recognize some deep structure to accommodate the productivity inherent in the linguistic data.

In other words, it is important to prove the sufficiency of a simulation to deal with regularities in the data, and demonstrating this can be a significant methodological result—especially if a given model was previously considered to be too weak to meet this condition (Hunt 1989, 608). Hence the importance of Rumelhart and McClelland's (1986) connectionist model of past-tense acquisition (earlier arguments favored symbolic processing). Similar sufficiency considerations gave rise to hypotheses about the functional architecture of the mind—including, for instance, Fodor's (1983) modularity claim.

It must be stressed, however, that sufficiency considerations rarely if ever pick out a single model that one can adopt as the correct representation of a given phenomenon; rather, after ruling out insufficient models, one is still left with a number of mutually incompatible theoretical proposals which differ from each other in nontrivial ways (Hunt 1989, 608). There is a substantial difference between the class of all possible implementations of a certain algorithm that alphabetizes a list of words and the concrete algorithm that is actually implemented in my word-processing application. The gap between a sufficient model and a necessary one is wide (Simon 1979a, 373).

A more restricted approach to model building is to decompose the main computational process into distinct stages. If a weakly equivalent model is a sort of black box, this strategy consists in analyzing the black box into further black boxes that are tested for input/output equivalence one by one. This approach is adopted in the first two examples introduced in chapter 1; Newell and Simon (1972) break down the process of solving cryptarithmetic tasks into individual steps, and Rumelhart and McClelland (1986), similarly, model individual phases of past-tense acquisition. In general, the more steps the overall process has, the more input/output pairs the computational model has to include; therefore, there is less room for variation among the models. An important research strategy is to find ways to ensure sufficient temporal density (or resolution) of empirical data for inclusion in the computational model (Simon 1979a, 364). That is one of the reasons why verbal protocols, which are never of the required density, are not the only source of empirical data for testing; eye-fixation data offer much better temporal density for testing the cognitive simulation models built by Newell and Simon.

Testing whether a model corresponds to the process that creates the output data is much harder than establishing weak equivalence; this is true even if we consider the strongest possible version of it, which requires decomposing the computation and testing whether parts of the model correspond to parts of the process. Again, in Fodor's (1968) terminology, models that are tested on the process level are described as "strongly equivalent" to the phenomena they simulate.

A traditional way of testing the details of a process in psychology is to use chronometric methods—that is, reaction or response time. The additive-factors method (Sternberg 1969) is the most widely accepted such approach (for a review of how these methods function in psychology, see Meyer et al. 1988). This is not to say that the relationship between reaction times and algorithmic complexity in computer science is straightforward: complexity expressed in terms of big-O notation is not necessarily neatly translatable into real-time differences between various machines and algorithms. This is because when input data sets are of limited length, it might be hard to empirically establish significant differences between algorithms of different time complexity—not to mention space complexity (i.e., requiring different amounts of memory). Moreover, different algorithms may fall into the same complexity class, so establishing the difference between them in terms of the amount of time that the processing takes would hardly be possible. In other words, testing performance in time gives only rough estimates of the underlying process. But a rough estimate is better than none at all!

In more recent research, the theoretical posits of a computational explanation are often tested against neuroscientific evidence: if one can find some stable activation in the brain that corresponds to a category of tasks, the category can be realistically defined in terms of brain functioning (this does not invalidate mental chronometry, though; see Posner 2005). However, it is far from clear how to devise well-defined empirical metrics on the basis of this kind of evidence; this is especially true because of the low resolution of many brain-imaging technologies, which makes them unsuitable for warranting detailed claims about information processing in the brain:

> Because it is a major principle of brain function that information is carried by the spiking of individual neurons each built to carry as independent information as possible from the other neurons, and because brain computation relies on distributed representations for generalization, completion, maintaining a memory, etc. . . . methods that average across many let alone hundreds of thousands of neurons will never reveal how information is actually encoded in the brain. (Rolls and Treves 2011, 484)

And, for ethical reasons, we usually don't directly intervene into human brains to record the activity of individual cells. In animal research, however, neurophysiological interventions, ranging from ablations to single-cell recordings, might be a viable way to create empirical metrics. As humans share a lot of brain structure with other primates, it is possible to extrapolate the results of such studies to humans.

While the goal of classical models was weak equivalence, which is why they usually settled for mechanisms sufficient to accomplish selected cognitive tasks, modern cognitive modeling often strives for strong equivalence by accommodating psychological and neuroscientific evidence. The model of rat navigation proposed by Conklin and Eliasmith (2005) has been designed with an eye to facilitating empirical testing; its theoretical entities are all observable. Similarly, the robotic model of cricket phonotaxis proposed by Webb (2008) relies on detailed information about cricket anatomy. In short, these models go beyond weak equivalence—usually at the price of having to increase complexity and conduct further investigations. Criticizing them for focusing too much on the neural details of the mechanisms they represent seems completely off the mark. Alas, this is precisely what classical functionalist theories of explanation in cognitive science appear to imply.

As Baars (1986) notes, the cognitive approach in general rejects the extreme parsimony of behaviorism, but the price to pay is that theories are much harder to test empirically. Creating as many independent, observable parameters as required for computational explanations is a difficult challenge for cognitive science but one definitely worth taking up (for a modern introduction to the testing of cognitive models, see Busemeyer and Diederich 2010). Accounts of explanation differ in terms of how they view empirical adequacy and its evaluation, and it seems that a detailed story about theory evaluation cannot be independent of an account of explanation. To say what explanatory value is, one needs to say what explanation is.

3   Are Algorithms Explanatory?

The classical account of explanation is given in terms of the covering law (CL) by Hempel and Oppenheim (1948). A statement describing the phenomenon to be explained, or explanandum, should follow logically from the explanans, which is the statement of antecedent conditions and general laws (covering laws). In other words, the structure of explanation is that of a logical argument. An explanation is correct only if the argument is valid and sound. Moreover, this account brings explanation and prediction together: in prediction, we infer what is going to happen from the antecedent conditions and general laws; in explanation, we infer that the explanandum phenomenon is an instance of a regularity by finding the appropriate general law and ascertaining that the law's antecedent has been satisfied.

The simplicity of this construal is its greatest strength, and the source of its most serious troubles. It seems that all subsequent attempts to formulate an account of explanation have tried to improve upon this model by introducing explicit requirements where the original proponents of the CL account had left unspecified, pragmatic background considerations (for a classical review of criticisms, see Ruben 1992, chapter IV).

The CL model is simply too broad. For example, on the CL account, adding an irrelevant factor to the general law would not affect its explanatory power as long as it preserved the law's truth value. If, for example, salt dissolves in water, then it must also dissolve in *hexed* water; yet an explanation of the dissolution of a sample of salt by appeal to the fact that it has been put in hexed water suggests that hexing must have, at the very least, contributed something to the process. A similar problem is that we can omit relevant factors and still end up with a sound argument if the conclusion happens to be true for some other reason. Surely, we can infer that Marcin is dead from the premises that Marcin has ingested an appropriate quantity of arsenic and that whoever ingests such a quantity of arsenic dies. Alas, I could be dead because someone shot me in the head before I digested the arsenic. But the argument is still valid and sound.

It could be argued, however, that questions pertaining to the relevance of the premises in the argument are of a pragmatic nature and could be easily tackled on a case-by-case basis without making the whole account more complex.[1] After all, we are interested in a parsimonious theory of scientific explanation, and the price for parsimony might be that it is slightly too broad if unaccompanied by quite trivial pragmatic considerations. These considerations are, simply, what approaches based on logical inference idealize from.

When applied to computer modeling in cognitive science, the CL account would be most naturally rendered as a theory that treats regularities in information processing as general laws (e.g., Rapaport 1998). In this interpretation, the structures in which information processing takes place would play the role of antecedent conditions. And, indeed, there are theories that are naturally construed as providing algorithms of explanatory value; for example, the Hebbian modification rule in computational neuroscience can be taken to explain learning in this way (for a review, see Brown, Kairiss, and Keenan 1990). The same goes for David Marr and Tomaso Poggio's proposal to investigate algorithms that compute stereo disparity (Marr and Poggio 1976), and Marr's own focus on algorithmic explanations might be interpreted in a similar vein. But none of the four cases introduced in chapter 1 appealed explicitly to algorithms as laws; only by paraphrasing them, we might interpret them as instances of the CL explanations.

The rub is that laws of nature are usually regarded as universally binding, unrestricted to any specific period in time or region in space, whereas, plausibly, algorithms running on a computer in my office resemble non-exceptionless regularities rather than universal laws. Moreover, it seems that, given the physical fragility of computers, many background conditions must probably hold at any given moment for them to work. Perhaps, then, algorithms are *ceteris paribus* laws rather than universal generalizations (if there are such laws at all; see Woodward 2002 for a denial that there can be any exceptions to laws). Assuming we agree that physically realized algorithms are ceteris paribus laws, we might ask what makes an abstract mathematical entity, such as an algorithm, a law of nature. There are, I should expect, an infinite number of algorithms that might be constructed in computer science, but not all of them get implemented (e.g., some compute functions that are of no interest to anyone). If you defend the claim that algorithms are laws of nature, you will probably want to know the status of these unimplemented algorithms. So, are they dispositional or actual laws? If you choose to drape yourself on the first horn of this dilemma, you must, presumably, come up with a plausible general account of necessary dispositional connections in nature (in which case, I wish you good luck); opting, on the other hand, for the second horn would commit you to asserting the existence of an infinite number of laws that are not doing any actual causal work (or so it would seem, since contingent isomorphisms between chaotic fluctuations of physical systems and these algorithms do not count as their instances).

One way I can see to get out of this mess would be to suppose that algorithms reduce to something more basic. Assuming that this might be so, I guess computability theory would be a natural place to start looking for a viable reduction base: the universal Turing machine might fill the bill, or perhaps the abstract state machine (ASM), if you prefer to accommodate nonclassical algorithms (Gurevich 1995; for an extension of ASMs to analog algorithms, see Bournez and Derschowitz, forthcoming). The covering law would then be a description of the underlying machine, while the antecedent conditions would specify the machine's configuration. This appears to be the most parsimonious way of dealing with the abundance of algorithms *qua* laws.

Note, by the way, that the algorithms themselves might be very simple. This is especially true of connectionist networks, which are even sometimes said not to implement algorithms at all—an obvious conceptual confusion, as most of them execute effective procedures describable in terms of a Turing machine. If the networks rely on continuous values, the values are Turing-machine computable to the degree of precision specific to the hardware (for a more faithful representation of the algorithm, I would recommend ASMs again; remember also that perfect measurement of continuous values would be required for hypothetical hypercomputational networks, cf. Siegelmann 1994). What is meant in such a case is that elementary steps of a computation, and even the computation as a whole, are better described in terms of the configuration of the network than in terms of the operations it performs. The operations are always roughly the same; it is the weights in the network that are important in the description of the process. This constitutes a second reason not to treat algorithms directly as laws, but to model them via machines (in this case, simply as connectionist neural networks). It seems, therefore, that not only parsimony with regard to the number of laws posited but also some "naturalness" of description underwrites the decision to explain computation via laws relating to machines.

Even if we can plausibly treat algorithms (or abstract machine descriptions) as laws of nature of some sort, there are further problems with the CL account of computational processes. First of all, because of its generic nature, the CL account is simply not specific enough to be normatively adequate. In particular, it is not geared toward the special sciences, which posit fragile complex systems as theoretical entities, and the norms regarding the application of the CL to such cases are treated as merely pragmatic. But they surely go beyond mere utility. Adding further constraints on implementation of computation, as described in chapter 2, is not required by the principles of the CL explanation. Accounting for multilevel structures of mechanisms, which is characteristic of today's research agendas as shown in the two examples from chapter 1 (Chris Eliasmith's NEF and Barbara Webb's artificial crickets), seems to be actually inessential for explanation according to the CL framework. And yet it seems that researchers believe that realistic multilevel modeling of mechanisms is of a greater value than a mere proof that there exists a similar algorithm for another machine.

One could argue that there is a version of the CL account that deals with pragmatic considerations used in computational explanations— namely, the semantic view (Hardcastle 1996). While I largely agree with Hardcastle's stress on pragmatic contexts, I do not see how "semantic ascent" would make much difference to problems of the explanatory value of computational models. Hardcastle simply assumes that we don't describe systems that merely satisfy computational descriptions, such as tornadoes, as computational; she moves the burden to pragmatic considerations. This is hardly persuasive. We simply don't describe such systems as computational because such descriptions are not adequate by virtue of the causal structure of tornadoes. All the same, I speak of models and descriptions below when it is handy (e.g., where idealization needs to be accentuated).

As philosophical questions about the exact character of laws of nature are far from decided, I will assume for the sake of argument that there is a way to overcome problems with the restricted scope and fragile character of laws regarding machines; I will also grant that ceteris paribus laws are perfectly explanatory. An antirealist about laws, for example, might happily admit that algorithms or descriptions of machines are laws and treat them as instrumentally as any other law. A realist, on the other hand, might construe algorithms (or machine descriptions) as relations between universals, and so forth. I simply grant as plausible the point that we can explain how a computer performs a computation by pointing to an algorithm (or the configuration of an abstract machine).

There is a deeper problem with the CL account: complex phenomena rely on complex entities, and regularities that govern the behavior of complex entities are best understood against the background of what such entities are composed of. A generic account of explanation, such as the CL model, can hardly do justice to this. But in cognitive science, and in large modeling projects in general, we are rarely interested in simple entities. And, indeed, it could have barely escaped your notice that the most parsimonious representation of algorithms that is able to pin down the regularities inherent in them relied on proposing an underlying machine, even if the machine itself was an abstract entity. However, this observation does not seem to be motivated by the explanatory framework as such; it is just that we know that there is a deeper level of generalization available.

One more thing deserves mention. By my lights, the often criticized fact that the CL account lumps prediction and explanation together (see, e.g., Cummins 2000) is hardly a vice. This feature of the received view of science is not at all obsolete (Douglas 2009). Predictions offered by theories serve to correct them, and they act as checks on explanations. In cognitive science, prediction is necessary to test the accuracy of computational models on data that were not used to train or build the models (this is why training data and testing data must be kept distinct). Without predictive power, such models could simply be closely fitted to the data and not general enough.

The predictive power of computational explanations can be tested to evaluate their empirical adequacy and compare them with other kinds of explanations. If the predictions offered by an algorithm proposed for the phenomenon can be achieved using a less complex, lower-level law, then the explanation itself might be spurious (see section 2 above for references regarding complexity metrics that are useful for evaluating parsimony claims).

4 The Functional Account of Explanation

A second model, offered explicitly as a replacement for the CL account, is Robert Cummins's functional conception of explanation (Cummins 2000). Cummins rejects the idea that explanation in psychology is subsumption under a law; instead, he suggests that the special sciences, especially psychology, are interested in effects. Consider, for example, the McGurk effect (MacDonald and McGurk 1978), which occurs when the visual perception of someone speaking clashes with auditory perception in specially crafted experiments: what we see may influence how we perceive speech, so that instead of "da" we hear "ga," and so forth. This effect is clearly a regularity (though culture-dependent, it seems, as in some communities, e.g., in Japan, people do not look at the speaker's lips; see Sekiyama and Tohkura 1991). As Cummins writes:

But no one thinks that the McGurk effect explains the data it subsumes. No one not in the grip of the DN model [=CL account] would suppose that one could explain why someone hears a consonant like the speaking mouth appears to make by appeal to the McGurk effect. That just *is* the McGurk effect. (Cummins 2000, 119)

Cummins also rejects the idea that prediction is to be the result of explanation; one might perfectly well explain a system without being able to control or predict it. We may understand the behavior of a system in an idealized fashion, but never be able to predict it due to its great complexity. And, vice versa: we can predict the state of a system without being able to understand it—for instance, we can forecast the tides with tide tables without having the faintest idea as to why the tides occur.

It might be argued that Cummins distorts the CL account in that he does not appreciate that a general law must be shorter than a description of pertinent experimental data couched in terms of individual measurements. Indeed, a set of measurement data is not a law, and neither is a verbal description of the McGurk effect, provided it merely fits the data.

What is described here is actually best understood as overfitting (directly redescribing the experimental data instead of finding the underlying generalization). At the same time, Cummins is right about the tendency in psychology to talk about effects. The notion of "law" is hardly in use in many special sciences.

According to Cummins's view, effects are to be explained as the exercises of various capacities. Capacities, in turn, are best understood in terms of "functional analysis" (Cummins 1975). A capacity, or a disposition, is analyzed, or decomposed, into a number of less problematic dispositions that jointly manifest themselves as the effect in question. This joint manifestation will be understood in terms of flowcharts or computer programs. Cummins claims that computational explanations are just top-down explanations of a system's capacity.

Before I go on, let me digress to prevent terminological confusion regarding top-down and bottom-up relationships. These notions have several meanings (Dennett 1998, chapter 16). In the first place, top-down vs. bottom-up relates to cognitive processing; in this sense, one may speak about top-down influences when referring to the kind of situation where a cognitive agent's expectations affect his or her perceptions. This was a hot topic with New Look psychology, forcefully attacked by Fodor (1983), who argued that lower-level perceptual processes are encapsulated from high-level information. Along these lines, Marr's theory of vision (Marr 1982) is a good example of a bottom-up approach. However, popular as it is, this is not the distinction Cummins has in mind. In another sense, the top-down/bottom-up distinction amounts to what Braitenberg (1984) called downhill invention and uphill analysis:

> It is pleasurable and easy to create little machines that do certain tricks. It is also quite easy to observe the full repertoire of behavior of these machines – even if it goes beyond what we had originally planned, as it often does. But it is much more difficult to start from the outside and to try to guess internal structure just from the observation of behavior. It is actually impossible in theory to determine exactly what the hidden mechanism is without opening the box, since there are always many different mechanisms with identical behavior. Quite apart from this, analysis is more difficult than invention in the sense in which, generally, induction takes more time to perform than deduction: in induction one has to search for the way, whereas in deduction one follows a straightforward path. (Braitenberg 1984, 20)

What Braitenberg stressed is important in all computational modeling: synthetic models are (usually) easier to understand, so building them helps to analyze real systems. A synthetic model is not constructed on the basis of how its individual parts are specified, but rather according to a general organizational blueprint; in this sense, it is "top-down" or downhill. Incidentally, there is an interesting exception to Braitenberg's "law" called Bonini's paradox: sometimes simulations are harder to understand than real systems, which is the case with some connectionist models (Dawson 2004, 17). This, again, is not what Cummins has in mind. He is instead making a point about explanation and advocating a special, functional way of analyzing real systems rather than endorsing the creation of artifacts. What really seems to underlie Cummins's use of top-down vs. bottom-up is the idea that a top-down theorist has to understand the task realized by a cognitive system before he or she can investigate its elementary components; adopting a bottom-up strategy, on the other hand, would consist of starting with, say, individual neurons and trying to "assemble" them into increasingly complex structures. In this sense, David Marr is a prominent exponent of the top-down approach (the focus of the next section). Accordingly, and somewhat confusingly, Marr's bottom-up methodology (in the first sense) is top-down in the third sense, and bottom-up in the second (he did not construct complete artificial systems, but analyzed natural ones).

Given that task analysis played an important role in all four examples I introduced in chapter 1, it is a plausible supposition that explanations have this form, as they clearly specify the cognitive task as the explanatory target. At the same time, some people interpret Cummins's idea in a somewhat stronger fashion. For example, Rusanen and Lappi (2007) claim that functionalist explanation is a kind of systemic explanation that shows how the system can have some capacities. They claim that cognitive competence constitutes both the explanans and the explanandum. But if this is the case, then a description of the McGurk effect (a kind of capacity that people have) would automatically account for it. This is exactly what Cummins criticizes as a disadvantage of the CL model, so if his account has this property then it fares no better than the CL view. Task analysis is not an explanation: it merely specifies in detail the cognitive capacity of a system—it does not show, however, *why* the system has a cognitive capacity. Because descriptions of cognitive tasks do not identify the relevant causal factors, they are not usually predictive of the system's behavior. If you see this as an advantage—rejoicing in how this story marks a sharp distinction between prediction and explanation—then it is hard to understand why.

A functionalist might reply that his theory is normative in character (O'Hara 1994); rather than offering a causal account, functionalism says what norms govern the practice of computational explanation. O'Hara claims that expert systems (or knowledge-based systems) in AI embody a theory of expertise, or cognitive competence, and that this competence is normatively explanatory of these systems in that they (should) perform the task in an optimal way. However, this is a nonstandard use of the notion of explanation; it seems that asserting that a system has a given cognitive capacity amounts to classifying the system as having this capacity rather than explaining anything. Apparently, what we have here boils down to a conflating of description of (idealized or rational) behavior and explanation. We still do not know what causes the system to behave this way rather than another; we have no clue as to why it embodies this expertise; nor do we know how the expertise is actually realized in humans if we only have a logical model of expertise (which is just logically sufficient for the expertise). Important as it is, task analysis itself is not explanatory.

Since task analysis is ubiquitous in cognitive science, it is fairly uncontroversial that it should feature in our account of computational explanation. Without it, we would not know what the explanatory target is. Yet there are some differences between modeling strategies. Connectionist models are usually developed cyclically, which is to say that only a partial specification of the task is ready before the first prototype is run. (This is why "probing" the model is important, as Cleeremans and French [1996] rightly remark.)

One cannot criticize most connectionist research for being bottom-up— that is, lacking clear explanatory targets (which are, admittedly, made clearer as they are developed). In this sense, connectionist models are usually top-down. Some of them are synthetic—but fail to be more understandable than their target phenomena. This failure, however, has nothing to do with their being bottom-up. The only sense in which they are bottom-up is that they were supposed to be more biologically plausible than classical AI, even if standard connectionist models are only remotely similar to real brains (see Bechtel and Abrahamsen 2002, chapter 10, for a review of how biologically implausible connectionist networks are). This plausibility, however, was a first step on the road to reduction, which was clearly the agenda of the Churchlands (Churchland P. S. 1989; Churchland P. M. 1996); reduction was seen as abominable to people like Fodor (1974), who feared that reduction would undermine the science of the mind and its autonomy. For them, top-down analysis is also linked with the traditional ideas of the autonomy of psychology. However, specifying the explanandum phenomenon does not commit one to disallowing reductionist explanation.

Moreover, the price to pay for psychological autonomy is steep, as it requires the discarding of all localization principles (such as the ones I proposed in chapter 2 to deal with difficulties found by Searle and Putnam). This, in turn, makes functionalism trivial (Godfrey-Smith 2008). For me, this price is too high, and I see it as a disadvantage for any theory of explanation. Let me elaborate on this.

Because of the (purported) autonomy of explanation, functional analysis cannot go beyond weak equivalence, and this is a serious drawback. In other words, we cannot satisfy Fodor's (1968) requirement of strong equivalence by following his own advice (Fodor 1974, 1975). The result of a decomposition is *an* analysis of the system into a number of subcapacities, but "what reason is there to think that that is how our brains do the job?" (Cummins 2000, 131). In section 3 above, I discussed a number of strategies adopted to reduce the set of candidate explanations. In Newell and Simon's (1972) task decomposition strategy, for example, the capacity to solve a cryptarithmetic problem is broken down into a number of subproblems. Yet nothing in the functionalist theory of explanation requires that the decompositions be realistic—the result being that no distinction at all can be drawn between weakly and strongly equivalent models.

Pace Cummins, even connectionism analyzes the capacities under investigation into more primitive elements, though it assumes certain dispositions to be basic. Accordingly, Cummins's charge that, allegedly, connectionism adopts a bottom-up strategy is clearly wrongheaded. Rumelhart and McClelland (1986) did not omit to perform task analysis, nor, indeed, did they construct their networks on the basis of neuroscientific evidence; as I observed in chapter 1, the model they proposed to account for past-tense acquisition was biologically *inspired* rather than biologically *correct*. It relied, for example, on the so-called Wickelphones, or specially devised representations of phonemes, which Rumelhart and McClelland had no neuroscientific grounds for introducing; they used Wickelphones because they needed a manageable phonological format sufficient to represent English verbs. It follows from this, incidentally, that what Rumelhart and McClelland performed was a top-down functional analysis in Braitenberg's sense: theirs was a completely synthetic model—the result of a downhill invention rather than uphill analysis.

A research project genuinely aiming at strong equivalence is Conklin and Eliasmith's (2005) account of rat navigation (for more details, see section 1.4). In this case, the model has been devised to reflect the latest neurophysiological evidence and has been shown to reproduce relevant, experimentally observed behavior. The researchers have also sketched a high-level mechanism responsible for navigation to the starting location of the rat based only on self-motion velocity commands. What this means is that, despite being a downhill invention, Conklin and Eliasmith's computational model (developed as an attractor neural network with spiking neurons) is the result of an uphill analysis. And that, in turn, shows clearly just how unilluminating the distinction between top-down and bottom-up strategies becomes when applied to more sophisticated modeling projects. Not only do these two types of strategies not exclude each other, but also the idea to adopt both at the same time has great appeal. After all, it seems only natural to exploit our knowledge about functionally construed cognitive capacities *and* our knowledge about their physical realizations so as to constrain the otherwise arbitrary decisions of the modeler.

The problem with Cummins's account is that Conklin and Eliasmith's (2005) explanation is hardly the result of *analysis*; their decomposition of the rat navigation system is based on what is *known* about its causal organization. And, admittedly, causal organization is a matter of empirical research rather than analytic decomposition. Indeed, for all its fancy-looking flowcharts and computer programs, Cummins's proposal exudes an unmistakable air of armchair theorizing.

Yet Cummins (1983) stresses that functional analysis does not result in a state-transition system, and that computation should not be understood causally but in terms of logical inference. He illustrates his point by drawing an analogy with computer programs. Explicating programs as symbolic transformations, Cummins writes:

Anyone who has ever written a computer program has engaged in interpretive analysis of an information-processing capacity, for the point of a programming problem is simply to analyze a complex symbol-manipulation capacity into an organized sequence of simpler symbol-manipulating capacities–viz., those specified by the elementary instructions of the programming language. Indeed, any problem of interpretive analysis can be thought of as a symbolic programming problem, since every functional analysis can be expressed in program or flow-chart form, and interpretive analysis specifies its capacities as symbolic transformations. (Cummins 1983, 35)

Although in modern programming, the use of flowcharts is limited (programmers working on programs larger than five thousand or ten thousand lines do not rely on detailed flowcharts, because they are too unwieldy for complex system analysis—see Parnas 1972, 1056), and there are more programming methodologies, all classical programs can indeed be reduced to string rewriting operations, so the point is well-taken. But Cummins goes further:

When an information-processing capacity is exercised, a symbol gets transformed. Often, though not always, to understand how the transformation is effected, we need to see the causal sequences leading from inputs to outputs as *computations*–i.e., as stages in step-by-step transformations of the symbols interpreting inputs into the symbols interpreting outputs. This is the explanatory value of interpretation: we understand a computational capacity when we see state transitions as computations. (Cummins 1983, 43)

This is somewhat baffling. In computer science, there are two basic ways of construing computation—as string rewriting and as state transitions. But we do not need to interpret state transitions as string rewriting to understand them as computations. On the one hand, a properly general account of computational models, such as Yuri Gurevich's abstract state machine, does rely on the notion of the computational system as a state-transition system. On the other, the universal Turing machine, arguably, is conceived as a machine that rewrites strings of symbols. So there is some confusion in Cummins's understanding of state transitions. This confusion leads him to an unfounded criticism of the CL construal of explanation as causal subsumption (this is already incorrect, as Hempel and Oppenheim did *not* claim that laws have to be causal) because, according to Cummins, one needs to interpret state transitions as string rewriting to see them as a computation, and this interpretation makes computation something that is not explainable causally:

Explanation of state transitions in S just won't resolve the perplexities generated by S's possession of a discursive capacity, or any other sort of information-processing capacity. A record of the physically specified state transitions of an adder, together with the relevant subsumption, would leave us baffled as to how the thing adds, even if the record included only the transitions and features that are interpretable as executions of the instructions of an addition algorithm. In reality, the record would not be so conveniently selective if it were constructed without the regard to its interpretability: from the point of view of physics . . . a de-interpretation of the algorithm represents a completely arbitrary selection of characteristics and events. This is why strictly "bottom up" strategies in cognitive psychology—i.e., strategies that proceed with neurophysiology unconstrained and uninformed by cognitive interpretability—have no serious chance of generating instantiations of cognitive capacities. To recommend such a strategy invariably amounts to changing the subject. (Cummins 1983, 62)

This argument is invalid, since one can interpret the state transitions directly in terms of a mechanistically adequate model of computation; approximation to such a model is always possible via the use of abstract state machines. Saying that we would not understand how ASMs work is completely unjustified; after all, this is a formalism regularly used to analyze computation and in practical settings (e.g., for checking specifications or verification of programs). And there is nothing in the causal account of computation that implies that it should be framed in terms of a bottom-up strategy in Cummins's sense. The causal decomposition of a system is usually the result of various interventions on a number of levels of its organization (Craver 2007).

This means that the notion of functional analysis is not really applicable to mechanistically adequate computational models that rely on state transitions, because it recommends, without proper justification, a reinterpretation of such models in terms of string rewriting. Nevertheless, we can weaken Cummins's proposal so as not to exclude causal explanations (and strive for more predictions to make models more testable).

On the one hand, the functional conception of explanation does render task decomposition as a theoretically important achievement—a point on which the CL account is completely silent. On the other hand, it is too broad to distinguish between a sufficient model and one that is actually implemented by the brain. Some computational explanations remain at the level of sufficient, or how-possibly, models, and Cummins's conception makes sense of them by providing a good understanding of classical modeling efforts, such as production systems built by Newell or connectionist networks trained by Rumelhart. Ditto for highly abstract Bayesian models, which are especially designed to be more "flexible" than connectionist networks and to ignore the lower-level details as "disturbing factors" (see, e.g., Griffiths et al. 2010).

On Cummins's functional conception, computational explanation should be conceived as necessarily idealizational. The rub, however, is that Galilean idealizational explanation, at least in Nowak's version (which I endorsed at the beginning of this chapter), implies the possibility of concretization, or removal of the distortions when relating the model to real-world phenomena. For some how-possibly models, notoriously, all concretizations are biologically implausible. For example, some Bayesian models posit computational power that exceeds the capabilities of the human brain (Jones and Love 2011). So these models are *not* methodologically kosher examples of idealization.

One might reply to my objection by saying that I have simply chosen a biased account of idealization. So let us suppose that Cummins's view licenses minimalist idealization, where the possibility of de-idealizing is not strictly required (Cartwright 1989). However, such an idealization should include the minimal core of a phenomenon; otherwise, a completely different model will be chosen instead (Hartmann 1998). It follows from this that sufficiency models, based solely on functional analysis rather than on realistic decomposition, are not minimalist idealizations either, since there is no guarantee that sufficient conditions will overlap with necessary conditions. In short, as minimalist idealizations, they are misconceived. So, there must be something wrong with Cummins' conception if it cannot distinguish between good and bad idealizations; even worse, it seems to license a wrong kind of idealization, one that is at odds with general theories of idealizational explanation in the philosophy of science.[2] One hypothesis is that the weakly equivalent accounts that posit black boxes are just descriptive and preliminary to the causal-mechanical stage of development in science (Mareschal and Thomas 2007).

To sum up, Cummins's brand of functional explanation licenses building empirically untestable models, conflates description with explanation, and is satisfied with mere possibility demonstrations that cannot even be considered correct idealizations—neither Galilean nor minimalist. Almost in no respect does this theory fare better than the traditional CL account. Its only advantage is that it stresses the importance of decomposition as a research strategy, but makes it a purely armchair enterprise of logical analysis. It is even more liberal than the CL model, and, for this reason, it is both descriptively and normatively inadequate—current practice in cognitive science is more stringent.

4.1   The Magical Number Three, Plus or Minus One

A popular way to make the functional account more robust is by introducing a hierarchy of explanatory levels. In the context of cognitive science, the most influential proposal of such a hierarchy is due to David Marr (Marr 1982; for extended philosophical analysis, see Kitcher 1988 and Shagrir 2010a). Marr stipulated that there are three levels at which a computational device must be explained in order for us to be able to understand it completely: (1) the computational level, (2) the level of representation and algorithm, and (3) the level of hardware implementation (Marr 1982, 24–25).

At the computational level, we ask what operations the system performs and why it performs them. Interestingly, the name Marr proposed for this level has proved confusing to some commentators (Shagrir 2010a, 494) who tend to understand it in terms of the semantic notions of knowledge or representation. What seems to be the case, however, is that, at this level, we simply assume that a device performs a task appropriately by carrying out a series of operations. We therefore need to identify the task in question and justify our explanatory strategy by ensuring that our specification mirrors the performance of the machine. In short, the term "computation," as Marr used it, refers to computational tasks and not to the transformation of semantically conceived representations. Indeed, it is hardly surprising that other terms for this level have been put forth to prevent misunderstanding, perhaps the most appropriate of which is Sterelny's (1990) "ecological level," which makes it clear that the justification of why the task is performed covers the relevant physical conditions of the machine's environment.

The level of representation and algorithm concerns the following questions: How can this computational task be implemented? What is the representation for the input and output, and what is the algorithm for the transformation? Note that we focus here on the formal features of the representation—the features that need to be decided to develop an algorithm in any programming language—rather than on whether the inputs represent anything in the world and, if so, how. The algorithm is correct when, given the same input as the machine, it performs the specified task.

Since the distinction between the computational level and the level of representation and algorithm amounts to the difference between *what* and *how* (Marr 1982, 28), the former level can be explicated as the level of competence (as understood by Chomsky) and the latter (along with the hardware level, which concerns the physical realization of the algorithm and representations) as the level of performance. The competence is described in a way that shows that it is appropriate for an entity, given the physical constraints, such as the environment that it finds itself in or its processing speed. In other words, the question of *why* the competence is appropriate belongs to Marr's first level (Shagrir 2010a).

Marr's levels are actually relatively autonomous levels of realization (Craver 2007, 218) rather than levels of composition. In other words, at all these levels, we speak about the same device, focusing on its different aspects, rather than about various entities at different levels that comprise the same device. There are multiple realizations of a given task, so Marr espouses the classical functionalist claim about the relative autonomy of levels, which is supposed to underwrite antireductionism. No wonder that functionalists embraced Marr's levels as well. The specification of the levels ignores the internal structure of the device. After all, this is just another, albeit systematic, functional decomposition.

To see why Marr's account became so influential, let us look again at my four case studies. Interestingly, Marr criticizes Simon and Newell's heuristics modeling paradigm by saying that "a heuristic program for carrying out some task was held to be a theory of that task, and the distinction between what a program did and how it did it was not taken seriously" (Marr 1982, 28). This, however, was no longer true of their later research. Before attempting to model people's performance, Newell and Simon analyzed the very nature of the task and the kinds of representations it could have required (Newell and Simon 1972; see also chapter 1.2), which would correspond, roughly, to describing the computational level and identifying how it might be constrained at the level of representation and algorithm. They then used verbal protocols and eye movements to individuate the steps of the task and related them to a simulation program— again, at the level of representation and algorithm. The level of hardware implementation was almost completely ignored, though it must be stressed that Newell and Simon complained that the neuropsychological data they had were simply too scarce to rely on (Newell and Simon 1972, 5). Marr himself dismisses Newell and Simon's modeling as mere gimmickry (Marr 1982, 347–348), because he thinks that when people do mental arithmetic, they actually do something more basic. Alas, his dismissal can hardly be justified within his own account of explanatory hierarchy.

Connectionist models can also be analyzed on three levels. The model of learning the past tense of English verbs (Rumelhart and McClelland 1986; chapter 1.3) specifies the task as learning how to inflect verbs; then, on the algorithmic level, it offers an algorithm and a specially crafted representation of the input and output that shows how it is solved. The choice of representation is justified by appeal to the computational level, where it is supposed that Wickelphones correspond to aspects of human competence. The implementation level is left out, although it is suggested that, because of its biological inspirations, the model could be mapped onto brain processes more easily than the symbolic models of classical AI.

There was considerable controversy over the question of at which level the connectionist models are supposed to explain cognition. Broadbent (1985), for example, contended that they were only at the level of implementation, so the difference would be just about the hardware and the algorithm might just as well be rule-based, while Rumelhart and McClelland (1985) denied that this was the case by saying that they model the phenomenon at the *algorithmic* level. In other words, Marr's methodology applies to connectionist models as well as it does to classical ones, though, initially, the former tend to rely on less detailed task analysis than the latter. (Task analysis is usually refined after probing the model, which is why Clark [1990] suggests that connectionism replaces level 1 with level 0.5—a partial description of the task.)

Let's look at the rat navigation model built in the NEF framework (Conklin and Eliasmith 2005; chapter 1.4). We begin by describing in appropriate detail the rat's navigational task (level 1), specifying, among other things, that the only source of information available to the rat is its own movements. (Were the rat able to use GPS and remember his previous position in terms of geographical coordinates, the task would have been different—possibly, much easier!) Yet we also look for constraints at the implementation level (the neural level) to discover the regularities at the level of representations and algorithm. Already the bottom level contains representations, but Marr does not disallow that. The only thing that would worry Marr is that top-level modeling is rejected and the levels are not autonomous; antireductionism is not presupposed in this methodology.

Webb's (2008) cricket phonotaxis model (chap. 1.5) is no different. Task analysis—what the cricket does and why it does it this way rather than another, and then a proposal for *how* it does it at the algorithmic level—is part of the model. The environment as well as the physical makeup of the agent (the hardware level) is investigated in minute detail in order to restrict arbitrariness at the computational level. Marr exploited features of the environment to impose constraints on algorithms, so this too is in keeping with his analysis. Initially, the neural network at level 2 was an idealized one, but it was later adjusted to subsequent experimental data.

There are other positions that fit Marr's methodology.[3] Newell, for instance, characterized the top level as the level of knowledge (1981), and, despite certain dissimilarities, the two views have much in common. The knowledge level is the level of competence, as in Marr's account, and it is highly abstract and idealized. Descriptions at this level are to appeal to rationality principles. Newell proposes more levels between the symbolic computation and the hardware implementation (the levels are sorted by their grain)—yet neuroscientific evidence is still somewhat neglected. Newell's ideas also correspond to Daniel Dennett's conception of stances (see Dennett 1987): the intentional stance bears a close resemblance to the knowledge level in that it operates on the highly idealized assumptions of rationality (so it is more idealized than Marr's computational level). The design stance focuses on the principles of the organization of the system; the physical stance focuses on its physical makeup.

Does the fact that all the four cases have an analysis in terms of Marr's three levels imply that his hierarchy is a good methodological tool? On the one hand, it is certainly ecumenical and, with enough good will, one could probably apply it to most models found in cognitive science (possibly because of Marr's influence as well); it also sits well with traditional functionalism and the symbolic models of classical cognitive science (Pylyshyn 1984, e.g., accepts a similar hierarchy of the semantic, syntactic, and physical levels) because it interprets the levels solely in terms of input–output mappings (Polger 2004). On the other hand, however, the fact that it does not require mapping causal interactions onto the computational level entails that even the best competence model might not lead to any *actual* decomposition of the system into parts, phases, steps, and so forth (Dennett 1987, 75–76).[4] Yet scientists seem to believe that the levels they discuss are not merely *façons de parler* dependent on the conceptual apparatus of a theory that happens to be employed (Bechtel 1994).

Let me elaborate. The problem is that the competence model is spelled out as a pure functional analysis of the task (breaking down the overall capacity into a number of subcapacities). Though one might stipulate that the goal of linguistics is to have a theory of linguistic competence understood in this way, it is far from clear that such a theory would actually explain all the relevant phenomena in human beings. The empirical evaluation of a conjecture about linguistic competence boils down to testing the correspondence of its input/output specification to real cognitive processes. They are analyzed in a serial or stepwise fashion without taking into account any intervening internal states, and possible interactions of the operations that contribute to the system capacity are not tested directly. This is because of the way functional analysis has been construed by Cummins (Cummins 1975; cf. Mundale and Bechtel 1996, 490). In short, this kind of functional analysis does not go beyond weak equivalence. Again, it is hard to avoid the charge that it conflates description of a cognitive capacity with its explanation. In this respect, Marr's proposal cannot be an improvement over Cummins's, for it just embraces the same idea.

According to Marr's methodology, models of lower levels are also empirically tested independently from each other, yet this independence, instead of contributing to the robustness of the theory of a computational system, might simply lead to positing entities that cannot be identified across levels. Worse still, the higher-level subcapacities might turn out to be purely conceptual and epiphenomenal—playing no actual causal role in the behavior of the system. There have been attempts to defend such higher-level accounts as "program explanations": according to Jackson and Pettit (1990), for example, higher-level properties, though not causally efficacious, are causally relevant. This, however, boils down to saying that a functional description tells us what the behavior of a system could have been like in contrast to its actual causal history. But causally relevant aspects brought out in a functional description are already accounted for in the interventionist theory of causation, which requires that one build a complete, counterfactual-supporting causal-structural model of the phenomenon in question. As Woodward (2008, 182) notes, in the interventionist view, causal efficacy is roughly the same as causal relevance. The problem is not that higher-level entities are suspect in themselves; on the contrary, Woodward (2008) and Raatikainen (2010) defend higher-level causation by pointing out that it would be difficult, if not impossible, to reconstruct the causal exclusion argument (see, e.g., Kim 1993) in interventionist terms; in contrast to Kim's view, interventionism does not imply the existence of a single basic level of causation. Rather, the problem is that there might be two distinct kinds of causal stories: one on the algorithmic level and the other on the hardware level, with many algorithmic stories weakly equivalent to what we know about the hardware level. In Marr's account, this potential lack of interlevel identities leads to no difficulties, even if incompatibility ensued. This is at best counterintuitive, for it would be natural to expect empirical testing to contribute to reducing interlevel discrepancies and increasing determinacy and not vice versa. For this reason, defending higher-level causation by appeal to program explanation (as in Ross and Spurrett 2004, 611) seems ill-advised. *Contra* Kim, I do not think there is anything especially worrying about higher-level causation. But Kim is right to say that positing multiple realizability might lead to epiphenomenality.

Let me explain. Not all counterfactual claims are explanatory, and some of the modal statements licensed by attempts to save the causal relevance of the higher level without its causal efficacy may be explanatorily irrelevant. For example, saying that it was also possible to kill Kennedy with an atomic bomb doesn't seem very explanatory about his assassination. Similarly, the possible use of a machine-translation engine, such as Moses, to translate a document does not explain how people would translate it, even if the result were the same. Surely, Moses is a different realization of translation, but this is why it does not have any causal role in traditional human translation. How-possibly explanations do not seem to be the best way to avoid the charge of epiphenomenalism.

Another problem for Marr's hierarchy is that the top level contains the context (the environment of the task) as a constraint, but the lower levels do not. So there is a shift of grain when going down the hierarchy but also a shift of context (McClamrock 1991). This, however, might raise problems with understanding just how the lower-level account is supposed to explain the facts referred to at level 1—namely, facts that confirm that some cognitive capacity is appropriate in a certain environment. For example, the cricket phonotaxis will work only in the land environment, because the cricket's eardrums work only for chirps audible in the air. There is no relationship with the features of the environment, or so it seems, at level 2. The chirps of a male cricket are not part of the female cricket's auditory system. This might give rise to discussions over whether Marr's hierarchy presupposes an individualist or an externalist approach to cognition (for more references, see Kitcher 1988). And, what's more troubling, level 1 cannot be properly said to be realized by (or supervene upon) level 2, because, inter alia, level 1 includes entities and relationships that are present only in the system's environment (i.e., they are not part of the system itself); the chirps of a male cricket do not supervene on the auditory system of a female cricket. To fix this, one would either have to embrace methodological solipsism, discarding everything but the female cricket at level 1—which would amount to rejecting statements to the effect that the female cricket realizes phonotaxis (it would just react to motion commands based on some perceptual signals)—or accommodate features of the environment at the lower levels. Otherwise, this hierarchy of realization is simply flawed; there are higher-level entities, such as chirps, that are not realized by anything at all in the hierarchy.

A smaller, though not insignificant, difficulty concerns the practice of defining the levels by recourse to the grain of the description. It is quite natural to expand such an explanatory scheme by introducing as many further divisions as a researcher sees fit (for a proposal to introduce level 1.5, see Peacocke 1986; O'Hara [1994] proposes even level 1.6). Newell, for instance, proposes five levels below level 1 (Newell 1990, 47). However, it soon transpires that, in this sense, the term "level" refers to an arbitrarily specified scale of objects captured by the description, which has little to do with what one would treat as *real levels of organization*. The upshot is that one would be hard-pressed to justify the stipulation that a full explanation of a given phenomenon must involve (at least) three levels. Why not two, or four, or five? Because three is a magic number? It seems there is something correct about Marr's idea of multilevel explanation, but that's only because we expect the levels to be real levels of organization rather than (merely) levels of detail. After all, he suggests that the levels are supposed to characterize a system in a particular way (as a computational capacity, algorithmically, as an implementation of this algorithm) and not just on a certain level of detail.

Marr's levels are relative, not absolute. As we saw, the studies I analyzed in chapter 1 did not jointly establish a single computational level; in each of the four cases, level 1 was different. This is unfortunate if we want to integrate explanations of multiple phenomena after deciding whether the algorithmic level of one model could be identified with an implementation level of another model; in such a case, making the distinctions absolute seems more appropriate (like in Newell's and Dennett's versions, which are both explicit in taking the personal level as their starting point; cf. also Sun, Coward, and Zenzen's [2005] four-level hierarchy). Revising Marr's hierarchy so as to accommodate this point is easy. What is harder is acknowledging that levels of organization in complex systems are usually nested—not necessarily in a linear fashion (McClamrock 1991, 191). Webb's crickets can be described on multiple levels of organization, and some of their parts can be decomposed hierarchically. A single multilevel, linear hierarchy of levels of grain is hard to use in such a case.

My remarks are not meant to suggest that Marr's idea is completely wrong. The core of it—that we need to include multiple levels in the explanation of computational processes—is also extremely important to my account of computation. What is missing is real mechanistic organization.

5   Mechanistic Explanation of Computational Processes

On the mechanistic account of explanation, computational processes are explained causally. This helps to alleviate some of the standard objections that were directed against the CL model. There are at least five constraints that make the mechanistic account different from it (after Craver 2007, 26):

(E1) Mere temporal sequences are not explanatory (temporal sequences),

(E2) causes explain effects and not vice versa (asymmetry),

(E3) causally independent effects of common causes do not explain one another (common cause),

(E4) causally irrelevant phenomena are not explanatory (relevance), and

(E5) causes need not make effects probable to explain them (improbable effects).

The CL account as such does not meet these conditions, which is what the standard counterexamples (like the hexed water in section 2) exploited.

Craver (2007) developed a detailed account of the norms of mechanistic explanation that ensures the fulfillment of these requirements.

However, contrary to what many proponents of a causal-mechanical account of explanation tend to say, there is a natural link between the CL account and the mechanistic account. I think the best way to look at mechanistic explanation is not as an alternative to the CL conception of explanation but as a more robust species of it. The mechanistic account provides additional criteria that guarantee a reliable connection between the explanans and the explanandum. Causal processes obtain only when there are regularities in nature, and it does not matter very much if we call these regularities laws of nature, real patterns, or mechanisms (for a debate on this topic, see Leuridan 2010 and Andersen 2011). As far as logical structure is concerned, mechanistic explanatory texts conform to the strictures of the CL model: a description of the explanandum phenomenon will be implied by a description of the mechanism, which includes the antecedent conditions (the start-up and termination conditions of the mechanism). The difference is that mechanists view their explanation not as a logical argument, or any kind of representation of reality, but as a process in the world. According to them, it is not representations of reality that explain why mechanisms have certain capacities but reality itself. This difference in attitude remains relatively unimportant for the practice and methodology of science.

The mechanistic account has close affinities with other views on explanation, too. A functional explanation, especially if accompanied by attempts to decompose the system, might be seen as a weaker version of it. In functional explanation, it is enough to present a how-possibly explanation of the phenomenon in question, which yields a weak equivalence of cognitive models. If the system suffices to perform the task, the task is explained. The mechanistic account, on the other hand, demands strong equivalence; an explanation is complete only if we know what actually performs the task (Craver 2007, 112). Functional decompositions might be used to form hypotheses, yet if we have a number of models such that each stipulates different internal subsystems, they are not considered equivalent when they differ in terms of causal dynamics. Only functional decompositions that correspond to structural and causal organization are explanatorily relevant in the mechanistic view (for a deeper analysis of the differences between the functional and the mechanistic account, see Piccinini and Craver 2011). Note that some brands of functionalism, in particular ones inspired by the semantic view of theories, already appreciated the need for causal relevance and biological realism (Hardcastle 1996).

This point is especially important in the case of the how-possibly models of cognitive science. It is not enough to say that there is a flowchart of the system that shows that the task could be implemented in this or that way; we also need to know if it was really implemented this way, on pain of stipulating causally irrelevant parts. In other words, the mechanistic explanation explicitly rules out possible epiphenomenal decompositions.

The norms of mechanistic explanation require that computational mechanisms be completely described. This means that they cannot contain any black boxes or questions marks; as an advocate of homuncular functionalism might say, all homunculi have to be discharged into completely mechanical subsystems whose working is understood. An incomplete description is called a *mechanism sketch* (Craver 2007, 113); a sketch contains gaps in the specification of the mechanisms, either in the list of parts of the mechanism or in the list of the interactions or operations of the parts. (This does not mean, however, that a complete explanation needs to bottom out at a fundamental physical level!) Such gaps, sometimes masked with placeholder terms like "process" or "representation," are usually, in time, filled with more details; a mechanistic model that is partially understood is called a *mechanism schema*. Functional models might start their life as mechanism sketches, and, for this reason, many extant computational explanations fail to explain cognitive phenomena completely. Also, some of the models—such as extremely computation-intensive Bayesian models of cognition—do not seem to have any complete mechanistic models having plausible causal decompositions. De-idealizing such models, or adding detail, will not make them more empirically valid, as they cannot be implemented by the human brain.

As I already mentioned in chapter 2, mechanisms are thought of as multilevel structures that are analyzed into three levels, just as in Marr's methodology. Yet these levels are levels of composition (Craver 2007, 188–195). In other words, the top level, or the contextual level of the mechanism, is the one that displays a capacity to be explained (the explanandum phenomenon). The isolated level contains parts of the whole system, but not the system itself. (Moving down the hierarchy is not just about increasing the resolution to accommodate more detail: each level features different entities.) The constitutive level is the bottom level of the mechanism, and it spells out the composition of the parts. These part-whole relationships are construed mereologically (with one important caveat: only proper parts count). This is why nothing can suddenly disappear from the hierarchy. As described at the top level, a mechanism is embedded in some context, but the lower level contains only entities that are part of the mechanism itself. The part-whole relationship makes it clear why, for example, no social context will be included in the explanation of an information-processing mechanism at the isolated level. There is no danger of idle discussion over whether this means that mechanistic explanation is internalist or externalist. If contextual effects prove important, they will be analyzed as parts of larger mechanisms and thereby included in the isolated level; if not, they won't. To see what an explanation of a phenomenon should include, one has to identify the boundaries of the mechanism and its causally relevant parts.

Another improvement over Marr's levels is the fact that the hierarchy may be nested in whatever way we want; we can analyze any constitutive level of one mechanism as a contextual level of another mechanism. This point would be appreciated by those who stipulate that lower-level structures, such as neurons, are themselves functional (Lycan 1987). There is not a single level of function and a single level of structure in this approach. What is required, however, is that the constitutive level bottom out, or be spelled out in terms that no longer refer merely to higher-level capacities (see chapter 2.5). In other words, only if there is a way to decompose such parts into some lower-level elements and explain how they interact causally (and not only on the isolated level!) is the mechanism completely and correctly described.

This three-level hierarchy may therefore replace the not-so-fortunate hierarchy of Marr, as it does not share its deficiencies, and it represents the decompositions as both structural and causally relevant. Yet there is an important difference with functional analysis. There is no level dedicated to competence only. The highly idealized competence level stipulated by Chomsky, Marr, Dennett, and Newell forms no part of mechanistic explanation at all. It may be used to create a mechanism sketch, but is not considered to be explanatory *of a physical system*. Competence models predict facts about the competence in question and can be tested empirically—this is beyond question. Moreover, competence might be explained functionally by analyzing it functionally as a flowchart and so forth. This is not the explanation of a *mechanism*, however; it is the explanation of the *task only* (a similar view, i.e., that competence in Chomsky's theory is a description of the function, and not an explanation of how it is made possible, is voiced by Johnson-Laird 1987).

For example, a model of translation competence might be created to describe, in an idealized fashion, what it is to translate a text from one language to another. This model, however, is not an explanation of psychological mechanisms that enable translation but a highly theoretical model of translation as such. As a psychological explanation, it might be postulating completely epiphenomenal mechanisms, or it might be quite trivial. If we confuse an explanation of a task with an explanation of a psychological capacity to perform the task, we have to treat competence models as incomplete explanations—as mechanism sketches.

In mechanistic explanation, there are no mechanisms as such; there are only mechanisms *of* something. In the case of a computational mechanism, this something is the task—the competence, or the mechanism's capacity. Such mechanisms are explained as having capacities, or serving a function, and not just as complex systems. This point was already appreciated by Stuart Kauffman in his classical paper on "articulation of parts explanation": there are multiple *views* of the system that one could take (Kauffman 1970). Under one view, a heart pumps blood; in another, it emits throbbing noises. A single physical system might be explained in multiple ways if there are many ways to look at it as performing functions. If all such explanations are causally relevant and correct according to the norms of mechanical explanation, all such views should be considered equally justified (Glennan 1996, 2002). This is how a mechanistic approach to explanation leads to explanatory pluralism.

This is not to say that anything goes. The explanandum phenomenon might be a fiction—in that case, the explanation fails (Craver 2007, 123). The phenomenon might also be incorrectly specified. For example, one might lump together capacities that should be explained separately (e.g., accounting for memory as a single phenomenon, rather than as several capacities, such as short-term and long-term memory, etc., would constitute such an error). Or one could differentiate things that should go together. To make a long story short, the phenomenon has to exist, be properly specified, and its precipitating, inhibition, modulation, and termination conditions should be understood. Various experimental (and natural) interventions are therefore required to specify the phenomenon correctly (Craver 2007, 123–128). If the phenomenon is merely part of a larger process, the explanation might also be partial, so proper generality is also called for.

The requirement that the organization of the mechanism—its parts, their operations and interactions—be real means that all constitutive mechanistic explanations are essentially multilevel. Moreover, conjectures about organization should be tested empirically: the mechanism is modeled causally, and causal interventions can be used to test whether the causal structure is as described in the model. For this reason, lower-level empirical evidence serves to confirm or reject hypotheses about the constitution of parts and their interactions. At the same time, such lower-level evidence is understood in light of the higher levels of organization.

In the context of cognitive science, this means that mechanistic explanation is neither purely top-down nor bottom-up. Top-down constraints are used to describe the explanandum phenomenon, but this description is not set in stone. For example, neurological deficits might suggest that a capacity that used to be treated as a single phenomenon is actually a set of similar capacities that break down in various circumstances. This means that neurological evidence as such is important for computational modeling of human cognitive capacities, and that its inclusion serves a purpose.[5]

Mechanistic explanation is reductionist—decomposition is a primary strategy of scientific reductionism (Bechtel and Richardson 1993)—but it does not make higher levels causally irrelevant (for a mechanistic model of higher-level causation, see Craver and Bechtel 2007). The explanandum phenomenon is not dissolved when we bring neurological evidence to bear on the mechanistic model. On the contrary, instead of falsifying higher-level descriptions, it is used to correct them and supply empirical evidence of their approximate truth.

Let me return to our four case studies from chapter 1 and briefly review how they would be evaluated in light of the mechanistic account of explanation. Newell and Simon's model of cryptarithmetic tasks (Newell and Simon 1972) is a mechanism sketch at best. Although the model has predictive power, it lacks the lower-level detail. One of the reasons was the fact that the evidence was not available at that time; another is that Newell and Simon considered the lower level to be less relevant. However, in building their model, they used several constraints, such as the capacity of the short-term memory and the number of operations per second. These constraints are partially on the level of the mechanism's components. Yet, because we lack any conjecture about how the higher-level cognitive process is realized by the brain, the model is not a complete explanation. This essential incompleteness might be one of the reasons why Marr regarded the theory as unilluminating (Marr 1982, 347–348).

Rumelhart and McClelland's (1986) model of past-tense acquisition also lacks the lower-level detail. Although the network is said to be neurologically plausible (just as the symbolic model of cryptarithmetic was plausible psychologically, owing to considerations regarding the capacity of short-term memory), plausibility is not the same thing as empirical confirmation. The model was not tested on the neurological level. (Another difficulty is that the psychological assumptions engendered in the model are not completely correct; see chapter 5.2 for other problems with the empirical validity of the model.) It is a how-plausibly model and a mechanism sketch only.

In the case of the NEF model, the situation is different. The rat navigation system has been explained on multiple levels (Conklin and Eliasmith 2005) and, though the decoders of information postulated by the model are not empirically testable (which might mean they are just useful fictions), the explanation seems to be at least a mechanism schema. From the methodological point of view, the decomposition is both functional and structural, and causal dynamics are taken into account.

The cricket phonotaxis model (Webb 2008) is also a mechanism schema. It spans multiple levels, uses evidence about the cricket's sensory systems, its neurological structure, and so forth, despite the conscious simplification of the artificial network used to model real neuronal organization. In most cases of explanation, we must abstract from some details for it to be practically useful. The schema is a clear-cut example of correct Galilean idealization. The distortions are gradually removed in this biorobotic research program, and, with their elimination, more and more predictions based on the behavior of the robots apply to real crickets.

Evaluating mechanistic models involves not only criteria that can be used for CL models, such as parsimony and generality (which can be used to test whether computational models offer anything over and above some lower-level explanations), but also additional criteria that stem from the multilevel structure of these explanations. The most obvious one is that the levels are understood in a clear way. What is important in this context is the question of whether an explanation in terms of computation will not be displaced by a neurological one. In my account, this is an open empirical question. To be more specific, we can talk of actual computation only when there is causally relevant information processing going on in the mechanism. If we can explain and predict the explanandum phenomenon on the basis of a more parsimonious model that does not refer to information or computation, then Occam's razor applies: computation becomes a superfluous fiction. In real computers, however, this is hardly the case. For example, a standard electronic computer cannot be accounted for on the electronic level without loss in generality; without knowing the logical roles of the electronic parts, we could not really explain the nature of the mechanism.

In the case of real, in contradistinction to putative, computational processes, the higher-level description in computational terms (as opposed to the description in terms of some other level of their organization) can be said to effectively screen off the lower-level description (Hardcastle 1996; de Jong 2003). Screening off is defined by de Jong thus: "Informally, a property is said to screen off another property if adding the latter does not improve prediction or explanation (more precisely, A screens off B from C if adding A makes B irrelevant for C [but not vice versa])" (de Jong 2003, 308). Yet it is definitely not a conceptual truth that some kind of description will screen off another; it needs to be tested in practice.

The explanatory target is also crucial. If someone wants to explain why IBM 7090 is six times faster than IBM 709 by appealing to the fact that the former was a transistorized version of the latter, then the computational description will *not* screen off the engineering description in terms of electronic component parts. But to explain why both computers can run exactly the same code, one may appeal to the same higher-level invariant generalization that is true of both. In these two cases, the relevant classes of contrast in causal explanation are different. In the first, the higher speed of computation in IBM 7090 in comparison to IBM 709 is caused by transistors and not by optimized code or any higher-level logical feature of this computer. In the second, the same code runs because there is an appropriate similarity of the organization of electronic components, which enables the CPU to run the same code, rather than because the two machines are built of identical component parts. In other words, the organization screens off the features of the individual component parts, and this is why different physical components are organized into a system that has exactly the same number of degrees of freedom.

The shared logic of IBM 7090 and IBM 709 is a result of the specifically engineered electronic organization of both computers. Thanks to this organization, some, but not all, degrees of freedom of individual component parts become irrelevant to predicting the behavior of the system, as some differences in the physical makeup cancel each other out. In this respect, the higher, computational level of both systems does not have a new emergent property capable of causing the logic of the software to be the same; it is rather the limitation of possible degrees of freedom that is warranted by imposing organizational constraints on the electronic parts. Note also that there is no need to appeal to the logic of both computers as a spooky "top-down cause" that makes them run the same code; rather, the organization of the transistors, capacitors, and other electronic parts makes the implementation of the logic possible. This is a relation of constitution rather than causation (Craver and Bechtel 2006); IBM 7090 is composed of transistors, capacitors, and so forth, and composition is a part-whole relationship. Accordingly, accepting the explanatory relevance of a computation does not commit one to a notion of top-down causation of the kind defended by Jackson and Pettit (1990). The higher-level mechanisms are spatiotemporal entities that may figure in causal explanations—at least under the interventionist account.

Note that a constitutive mechanistic explanation of a phenomenon does not explain the bottom level of a mechanism—that is, its constitutive parts and their organization. There might be an explanation of why they are organized this way and not another, and why they are what they are, but this is not part of the same explanation. Most importantly in this context, this means that one cannot explain the makeup of parts that constitute a computational system.

An example will illustrate this point. I'm using LanguageTool (a Java application) to proofread a document in English (see Miłkowski 2010). The algorithm implemented in Java on my laptop computer does not explain why the integrated circuit performs instructions as specified in the machine code for Intel processors. When knowing that Java code is being executed, I could only predict a certain high-level pattern of these processes. This is why I find talk about the top-down explanation of systems by appeal to their systemic computational capacities (Rusanen and Lappi 2007) to be mildly misleading. All we can explain and predict by referring to Java code is a general tendency of the system and not its constitution. This is not a "top-down" explanation: "down" is missing altogether, so it is just a "top" explanation (which is mechanistic all right, but, like an explanation of the movement of billiard balls, remains limited to a single level of the mechanism).

Even more interestingly, the Java code does not explain how the CPU interprets machine code instructions, which is, arguably, on a level of composition above the level of electronics. The CPU level is computational in its own right, but it is not the target of an explanation which has Java code as its isolated level. There is no explanatory symmetry between the levels; it is the lower (–1) level that can account for the isolated level and not vice versa.

Note also that the contextual level of a computer is not explained computationally either. A description of how a computer "behaves" in its environment is couched in generic causal terms and not in terms of information processing. Obviously, one computational mechanism can be a component of a greater computational system, in which case the former is described at the constitutive of level of the latter, with the latter serving as the contextual level for the former. Integrating such computational mechanisms may consist in specifying both of them in computational terms; however, the uppermost level of the larger mechanism, as well as the constitutive level of the submechanism, will still remain noncomputational in character.

For example, the behavior of my laptop computer, when put on grass, might be influenced by the environment it is in (e.g., by giving rise to a short circuit). This influence, however, will not be computational, and we cannot explain it computationally. In other words, the mechanistic account of computational explanation does not render it useful a priori; its usefulness depends on the explanatory target and the organization of the mechanism in question.

Note that in phony conceptual computers, such as Searle's wall, a pail of water, or the economy of Bolivia, the level of information processing is redundant (although, arguably, some of the capacities of Bolivia's economy may be explained in terms of information flow and information processing; the economic agents usually peruse information before acting). We cannot predict anything new about the wall on the basis of its putative computational description. All the model's predictions would be completely useless, and the parts that were ascribed any causal role in the putative computation would have to fulfill the strict criteria of causal explanation (Woodward 2002). It would be incredible, however, if parts created by logical operations on descriptions (see chapter 2, section 5) could actually be causally relevant in the ways required by the model of computation. And without causal relevance the computational descriptions used by Searle and Putnam are not explanations at all.

Let me sum up the criteria of empirical adequacy that can be used to evaluate mechanistic models (see Glennan 2005, 457). First of all, the information accepted and generated by a mechanism, or its information-processing behavior, should be predicted by the mechanism. (The larger the range of empirically detectable inputs for which we have adequate predictions, the better.) We may call this *behavioral adequacy*. Mechanistic adequacy involves such factors as the completeness of the mechanism: enumerating all the components, their localization; specifying their causal profiles; quantitatively adequate descriptions of interactions; spatiotemporal organization; submodels of the components, if they are mechanisms as well; and whether only the mechanism in question is responsible for a given behavior, or one should include others that run concurrently to it or are redundant.

6 Theoretical Unification

Although computer simulations have been used in psychology and cognitive science for a relatively long time, it is surprising that complete explanations of complex cognitive phenomena in terms of computer models are quite scarce, even if today there are many more models than in the early days of cognitive computer modeling. (As Hunt [1968] shows, in that period, there were many more artificial intelligence projects than explanations of psychological capabilities per se.) One of the obvious reasons is that it is notoriously hard to model a complex phenomenon; it is even harder to evaluate a model of a complex phenomenon if it is unclear what parameters should be tested. Actual models tend to be partial—they are microtheories (as Newell and Simon [1972] call their research results) rather than general explanations; the theoretical integration and unification of models remains a highly prized long-term goal. (Already Simon [1979] argues that general frameworks will help in this, and Newell 1990 is a locus classicus.) To build theories of cognitive architecture is therefore one of the goals that computational simulations are supposed to strive for. Another is the modeling of multiple cognitive processes as they interact—or as Forbus (2010) calls it, macromodeling. Allen Newell stressed the need for unified theories of cognition, and the need is still not satisfied today, even if there are lively research programs focusing on cognitive architectures such as Soar (initiated by Newell and his collaborators), CLARION (Sun 2002), ACT-R (Anderson 2007), or LIDA (Franklin and Patterson 2006).

This is why researchers regard theoretical unification—of cognitive phenomena or of theories—as one of the main functions of computer modeling (Cleeremans and French 1996). By way of comparison, Mendeleev's discovery of the periodic table of elements was one the major theoretical breakthroughs in chemistry, and it cannot be dismissed as simply a sorting of data that people already had access to; it was not just a taxonomy, for it allowed chemists to discover new elements and isotopes. Yet it was not explanatory of the elements themselves in any straightforward sense, notwithstanding its usefulness in systematically predicting a number of their properties.

It is quite natural to suppose that an account of explanation in terms of unification, advocated by Philip Kitcher (1989), might be applied to computational models; the more facts that can be explained, the better. A general model, or a class of models, that can be used to derive descriptions of many phenomena is therefore of explanatory value. Nevertheless, the unification account of explanation cannot be treated as undermining other explanatory frameworks in the philosophy of science, as it has its own problems. For example, it would assess an explanation of an uncommon phenomenon as less explanatory than that of a common one (Craver 2007, 47–48). This seems to clash with standard scientific practice, especially if the uncommon phenomenon is an anomaly for existing theories.

I submit, then, that unification, as a feature of modeling, is desired—though not strictly necessary for its explanatory value—if there is no mechanistic explanation at hand. (Glennan [2002, S532] argues that unification is genuinely explanatory when it refers to the higher-level structure of common mechanisms.) This is not to say that integration of theories is not important in cognitive science. On the contrary, I think that the goal formulated by Newell in 1990 is extremely important in theorizing about cognition. Recent proposals of integration via modeling the mechanisms in cognitive robotics (Morse et al. 2011; D'Mello and Franklin 2011) seem to be especially interesting from the mechanistic point of view. As D'Mello and Franklin argue, experimental psychology does not require complete models, as robotics does. In effect, they seem to follow the steps of early roboticists, such as Walter (1950), who stressed the importance of complete models, but their explanatory target is the human being rather than a behavioral capacity of an imaginary creature (e.g., Walter's artificial tortoise).

7 A Checklist for Computational Explanations

1. Is it clear what the explanandum phenomenon is?

The phenomenon to be explained should be clearly identified. Complex computational models might look nice, but they can also prompt a certain kind of computational fetishism; despite any intuitive persuasiveness it may have, an impressive computational gimmick without any clear explanatory target will not contribute to our understanding of the world. This is a trivial requirement, but it's worth noting that in order to evaluate a model we need to know its explanatory target.

2. Is the explanandum phenomenon analyzed and well understood?

The logical structure of the description of a phenomenon (e.g., a specification of the grammar structure) is needed for explanations to be complete.

3. Is the explanation general and does it predict previously unobserved behavior?

If not, the computational model might just be an instance of overfitting.

4. Are all algorithms specified?

Some predictions are possible even on the basis of incompletely specified algorithms, but comprehensive predictions require a full account; this is true even in the CL concept of explanation.

5. Is the model implemented?

Verbal descriptions or even more formal specifications of models are not as methodologically valuable as complete implementations.

6. Does the phenomenon to be explained essentially involve the processing of information?

If not, the explanation might be spurious. It could be a redescription of a physical process. There are two considerations here. First, the phenomenon must be substrate-neutral as an information-processing phenomenon. If it seems absurd that the phenomenon is substrate-neutral (e.g., earthquakes are substrate-dependent), then the description lacks explanatory value. Second, if there is a simpler noncomputational theory with comparable predictive power, then we have reason to believe the computational explanation to be redundant. A theory, as I've pointed out in several places in this book, can be evaluated as simpler than another theory on the basis of such metrics as minimal description length or Kolmogorov complexity.

7. Are the components of the system that displays the explanandum phenomenon specified?

The role of components in contributing to the explanandum phenomenon should be specified. This is what the functional account requires.

8. Does the explanation fulfill the mechanistic norms, and is it empirically adequate?

As required by the mechanistic account of explanation, the components should be completely specified on multiple levels. This is not a simple yes-or-no choice: some details of organization might be left unspecified if they will not contribute to higher accuracy of the model or affect its depth. Note that idealized explanations (mechanism sketches) are also valuable, but it should be possible, in principle, to fill in the gaps in the sketch.

Now, with these criteria in hand, we can see that tornadoes, pails of water, or walls do not seem to be usefully explainable in a computational way. First of all, it is hard to say what would be the explanandum phenomenon that they exhibit. But even if someone stipulates that they do what they do (e.g., cause damage to cities) because of the computations they perform, he or she needs to specify and analyze the phenomenon that is being explained. The problem with such explanations is that they would be easily overridden by simpler meteorological, physical, or architectural explanations of lower complexity (see question #6). This is already true according to the CL account of explanation, and it is even more obvious in the mechanistic approach.

8 Summary

In this chapter, I reviewed the role of computational models in the explanation of behavior of complex systems—cognitive systems included. I stressed that these models offer theoretical rigor, but they should be accompanied by a similar rigor in empirical testing. How the empirical tests of computational models would be analyzed depends on our conception of explanation itself. I have analyzed three conceptions of explanation: the covering-law (CL) model, the functional account, and the mechanistic account. Each successive account of explanation might be considered a more robust version of the preceding one. I argued that the mechanistic account has distinctive advantages over the functional one, though the latter is usually considered the standard in the philosophy of cognitive science.

It transpires, therefore, that as long as a computational explanation of an information-processing system proceeds according to the norms of mechanistic explanation, it can be related to other explanations of the system, each of which targets a different level of organization. If a computational explanation were completely autonomous, we could not rely on evidence from a variety of scientific disciplines, which would be unfortunate; if it were completely reducible to other kinds of explanation, it would be heuristically interesting at the most; the theoretical entities posited in such explanations could just as well turn out to be useful fictions.

4 Computation and Representation

My aim in this chapter is to show the explanatory role of representation in computational cognitive science. This is a highly debated and controversial topic, and for a very clear reason: for many, the computational theory of mind is vindicated because it makes place for intentionality, or representation, in a physical world (Fodor 1980; Newell 1980; Pylyshyn 1984). Jerry Fodor defended vigorously the representational character of computation, aptly summarized by his slogan "no computation without representation" (Fodor 1975, 34). But this is exactly what I denied in chapter 2. There are computations that have nothing to do with representation as it is commonly understood—as something that has reference and content. A Turing machine that just halts and does nothing else is a fine example of a nonrepresentational computation; a logical gate is another.

However, one could say that, being formally equivalent to a logical operation of propositional calculus, a logical gate—for instance, an AND gate—does represent something. Indeed, this is why AND gates are manufactured in the first place. So I need to show in what sense all computations are taken to be "representational" and why I believe this way of speaking amounts to stretching the term's meaning beyond its sphere of useful application.

At the same time, I want to apply the general account of computational explanation to cognitive science. Representation has a role to play in cognitive science, and computational models may rely on representations. Cognitive systems use information to get by in the world; they process information, and, under special conditions, the information they manipulate becomes representation.

So my task is first to uncover the motivations behind the claim that there is no computation without representation, and then to dispel confusion surrounding the notion of "symbol," which is sometimes used interchangeably with "representation." I will also show some fatal flaws that the classical views on representation, both symbolic and informational, have. Next, I will sketch an alternative conception of representation that will be useful in analyzing representational explanations. As in the previous chapter, I will rely on the mechanistic model of explanation, framing my account as an abstract specification of representational mechanisms. Finally, I will turn to an analysis of my four cases introduced in chapter 1. The limitations of space do not permit the inclusion of a full discussion of all the corollaries of my view, and I do not wish my model of representational mechanism to compete with book-length theories of representation. This chapter is supposed merely to capture key organizational requirements for representation, not to explain everything in detail.

The upshot of my discussion will be that there is something to the Fodorian slogan if you have it backward. No representation without computation. But surely a lot of computation without representation.

1 Symbols in the Physical World

One of the advantages of the computational theory of mind was that it held out a promise to explain mental representation in a physical world. In short, computation was supposed to naturalize intentionality. For example, Allen Newell writes:

> This concept of a broad class of systems that is capable of having and manipulating symbols, yet is also realizable within our physical universe, has emerged from our growing experience and analysis of the computer and how to program it to perform intellectual and perceptual tasks. (Newell 1980, 136)

Newell was not alone; Pylyshyn (1984) makes much the same general point, and a similar argument can be found in Craik (1943, 57), not to mention other early works in cognitive science (notably Miller, Galanter, and Pribram 1967, 50). But was all that optimism well founded?

Before I turn to this question, an important reminder: the scientists' programmatic manifestos and philosophical reflections have a different status than real scientific achievements. For one thing, such pronouncements usually describe science at a high level of abstraction; they draw on genuine scientific results to speculate about various breakthroughs "waiting just around the corner" and, often, to convey ambitious visions of future progress. They might be expressions of methodological false consciousness or hyperbolic promotional material aimed to make a research program appear special. Either way, a naturalized philosophy of science does not have to take such enunciations for granted, especially if they announce the beginning of yet another scientific revolution; accordingly, I will not treat them as truths set in stone.

The methodology embodied in a research program might actually be different from what is advertised in an introduction or programmatic manifesto. The actual methodology is what matters most; methodological consciousness comes second. In other words, the philosophical views of David Marr, their interest for students of philosophy notwithstanding, are of secondary importance when dealing with metatheoretical issues. I mention Marr because philosophers hoped to resolve the dispute between internalism and externalism by looking at his work, and they did not distinguish between its metamethodological content and the actual results. I do not wish to follow the same path when reading Newell; his philosophical views are as debatable as any, but they are not on a par with his results in computational modeling of cognition.

Let me return to the question: was the enthusiasm regarding the computational theory of mind's potential to explain mental representation in a physical world well founded? In effect, my answer will be partly negative and partly positive. No, the argument offered by Newell (and Pylyshyn) cannot work, as it relies on a nasty equivocation. Yes, computational theories help in explaining representation, but the connection between the two is much more complex than is usually supposed.

It is instructive to read two more sentences from the passage I quoted: "The notion of symbol that it defines is internal to this concept of a system. Thus, it is a hypothesis that these symbols are in fact the same symbols that we humans have and use every day of our lives" (Newell 1980, 136). No grounds are given to substantiate this, although, being an identity claim, the hypothesis can be easily disproved, as I will shortly demonstrate. Worse still, because Newell's claim appealed to two distinct notions of symbol, the hypothesis gave rise to a serious conceptual confusion within cognitive science, leading to more trouble later, as yet another (third) sense of the term "symbol" came into play. Instead of a valid argument for a computational theory of mind, we have an equivocation. And there are discussions of the role of symbols in cognition in which it is never really clear which of the senses is employed.

Thus, it is advisable to tread cautiously. In computer science, the word "symbol" functions in at least two different meanings, which Newell seems to conflate at times. In the first meaning, a symbol is simply a token from an alphabet, or a piece of information that is processed by a computation. For example, it is customary to talk of symbols on the tape of a Turing machine (e.g., Cutland 1980, 53–54). However, these symbols are purely formal. They need not possess any content; they need not refer to anything. Yet if you conflate representation with formal symbols (and take into account that any computation operates on formal symbols [i.e., pieces of information that are processed]), then the claim "there is no computation without representation" will come out true—trivially true.

It is, of course, possible to ascribe some meaning to formal tokens, but if you just pick one meaning from the list of all possible models that satisfy the description of the computation, this ascription will be arbitrary. The ascription is effected from outside of the computational system. This might be seen as an advantage that allows greater universality, but not all human representations are arbitrary in this way (Touretzky and Pomerleau 1994).

At times, Newell and Simon suggested that this is *the* meaning of "symbol" that they had in mind. For example, Vera and Simon defended the claim that connectionist networks used symbols in this sense (Vera and Simon 1993). Note that formal symbols in digital systems do not admit of anything that could be a *subsymbolic* component—so often brought up in the context of artificial neural networks. Subsymbolic states of a Turing machine tape have no bearing on the processing.

A second meaning of "symbol" draws on LISP, a programming language heavily employed in classical AI. In LISP, a symbol is a pointer to a list structure (Steels 2008, 228). The structure contains the symbol's name, temporarily assigned value, a definition of function associated with this symbol, and the like. Note that Newell (1980) often mentions LISP, and his insistence on access and assign operations as essential to symbols definitely gives away his reliance on LISPish ways of thinking.

On this second meaning, the physical-symbol hypothesis makes a bit more sense, as there is at least some hint of content in the LISP symbols. Being pointers, they are used to access information. Alas, what they point to are other formal symbols. Newell and Simon seem to rely on this reading of symbols in several places:

What makes symbols symbolic is their ability to designate—i.e., to have a referent. This means that an information process can take a symbol (more precisely, a symbol token) as input and gain access to the referenced object to affect it or be affected by it in some way—to read it, modify it, build a new structure with it, and so on. . . . In discussing linguistic matters one normally takes as prototypic of designation the relation between a proper name and the object named—e.g., George Washington and a particular man who was once President of the United States. . . . In our theory of information processing systems, the prototypic designatory relationship is between a symbol and a symbol structure. Thus, X2 is the name of (i.e., designates) the list (A, B, C); so that given a symbol token X2 one can obtain access to the list—for example, obtain its first element, which is A. (Newell and Simon 1972, 24)

The peculiarity of this notion of designation is clear. (There is a formal definition in Newell 1980, but it suffers from other fatal flaws, as it is too broad even if one accepts its peculiarity; see Ramsey 2007, 9). These symbols are still just formal; no referent in the outside world is given. However, Newell sometimes mentions the external environment:

The system behaves in interaction with the environment, but this is accounted for entirely by the operation of the **input** and **behave** operators. The operation of these two operators depends on the total environment in which the system is embedded. (Newell 1980, 147)

The input operator in his system architecture creates new symbolic expressions, and there is no mechanism posited that could autonomously generate expressions that refer to the external world. In other words, the meaning of these symbols is still parasitic upon the observer. They are formal per se, even if they are not so broadly construed as to cover all possible computational models.

Alas, LISP is a higher-level programming language, which means that it can be implemented in various ways depending on what kind of processing architecture it runs on. More specifically, a higher-level symbol might get translated to assembly instructions that bear no resemblance to any access operations that Newell mentions (for more details, see Scheutz 2000). In other words, the observer must also interpret the system in a special way to see the formal symbol at a given level of abstraction. Interestingly, an artificial neural network might implement a LISP interpreter; if you wish, the states of the network might then be subsymbolic. In fact, Newell hoped that neuroscience would find out which physical architecture supports a symbol system, so he might have embraced connectionist systems as a step in that direction (Newell 1980, 174).

Let us now compare these two notions of a symbol with "symbols that we humans have and use every day of our lives" (ibid., 136). Does Newell mean public symbols, such as letters from the Latin alphabet? Or does he mean that mental representations are symbols? In the first case, the identity would be quite plausible. We do use symbols that are arbitrarily assigned meaning. Public symbols help us in cognitive processes—as the proponents of embodied, embedded, extended, and enacted (4E) cognition tirelessly remind us (e.g., Clark and Chalmers 1998; Wilson 2004; Wheeler 2005; Clark 2008). Interestingly, Newell and Simon stress the importance of the environment in which humans are embedded, and they hypothesize that the task environment shapes the search space (of representations) that humans use to solve problems (Newell and Simon 1972, 789–790). One could also point out that tracking the subjects' eye movements while they solved cryptarithmetic tasks was motivated by the realization that the symbols they were operating on were *external*. Read this way, Newell and Simon would be, surprising as it may seem, the forerunners of the 4E movement. Yet their hypothesis might mean something else:

Stated another way, the hypothesis is that humans are instances of physical symbol systems, and, by virtue of this, mind enters into the physical universe. (Newell 1980, 136)

This statement could be understood in a way that is compatible with Fodor's methodological solipsism (Fodor 1980): people manipulate mental symbols. These symbols are formal in that they point to other formal symbols, but this is the best theory of cognition available. That symbols merely designate other symbols would be quite obvious for (early) Fodor. The symbols that are at the beginning of the story about what other symbols mean would be the symbols from the lexicon of the language of thought (Fodor 1975). Note that this use of the term "symbol"—referring to an element of the language of thought or, roughly speaking, to some linguistic entity bearing the meaning of an individual concept—gave rise to yet another meaning of this term. In later discussions, connectionists applied the term "subsymbolic" to something representational at a level below that of the concept-like symbols of Mentalese.

Although turning designation into a symbol-to-symbol relationship has an air of paradox to it, it was argued that the overall conception has some advantages for modeling cognition. Arbitrary symbols can be syntactically rich, and that could allow for their compositionality, which is what Fodor thinks is a necessary feature of any theory of representation flexible enough to accommodate properties, such as productivity, of the human linguistic capacity (the locus classicus of this line of reasoning is Fodor and Pylyshyn 1988). Newell and Simon advanced similar arguments by linking the arbitrariness of symbols with the universality of the physical symbol system (Newell and Simon 1976). As Newell stressed, flexibility of behavior is central to universality (Newell 1980, 147). In the Newell and Simon version, the argument for universality was based on the assumption that a machine that can use symbols to designate in arbitrary ways will also be able to describe, or code, another machine, and this is the core of the universality of any computational model. But it may be contested whether compositionality, systematicity, or universality really are properties of human behavior. That behavior may be fruitfully described in these terms is beyond doubt; yet empirical validation of such descriptions is not at all straightforward, especially in the case of systematicity, which was introduced only informally (for a review of arguments from the systematicity of behavior, see Aizawa 2003). Note also that while connectionists seem to accept the claim that the symbols of Mentalese are structured (for a brief review of how connectionists reply to Fodor and Pylyshyn, see Bechtel and Abrahamsen 2002, chapter 6), some philosophers deny that symbols have any structure at all (Cummins 1996) or claim that symbolic structures are less appropriate for cognition than distributed connectionist representations (Chalmers 1993).

To sum up, even if symbolic theories of representation are correct, and representations are best conceived of as symbolic, it is not at all clear what advantages these theories bring. The claims—that they enable universality and flexibility of behavior—are quite strong, but it has not been empirically shown at all that people are always flexible. Also, it is not clear what systematicity actually amounts to; different theorists construe apparently simple features, such as the structural nature of symbols, differently. A proper defense of the symbolic approach to representation would require more precision; basic arguments cannot be so vague and ambiguous.

Another point worth stressing is that the syntactic features of the symbol system are available to the system. In other words, if manipulation of formal symbols is all there is to representation, then representations can be said to be available to an agent that has the capacity to compute or manipulate them. Under this proposal, the agent has access to a representational system of considerable richness. The availability of the representations *to* the system rather than to the external observer may be used to explain intensionality or referential opacity (Fodor 1975, 1980). This is an important point, to which I will return when developing my own account.

Nevertheless, all that the cognitive system has access to are the syntactic properties of symbols rather than their reference. This means that, for the system, the symbols constitute uninterpreted formal entities, which can acquire content and reference *only after* they have been interpreted. The rub is that, unlike utterances, mental representations do not seem to require any *external* interpretation to be meaningful. And if they did, we would have to explain external interpretation as well, which would lead to a vicious circle, because a theory of external interpretation would presumably appeal to the representational abilities of the observer. And if we disposed of the external observer and posited an internal homunculus instead, it too would remain undischarged: another homunculus would be necessary to interpret its representations, and so on. Because of this circularity, a conception of representation characterized in terms of formal symbols cannot really account for the emergence, or constitution, of representation.

But if, on account of the circularity it breeds, we reject the claim that any interpretation is required for mental representations, then the symbols will remain purely formal. Fodor (1980) went so far as to say that it is simply impossible to have a theory that specifies the reference and content as anything outside the cognitive system. Such a theory, according to Fodor, would need to include knowledge of what our representations *really* refer to—a knowledge we won't have until we arrive at complete science. Claims about reference depend on claims about causation (if you endorse a causal theory of reference), which, in turn, rely for content on natural laws. Yet knowledge of natural laws requires that we know all the natural kinds that our terms refer to. We need to know them in order to know the causal factors, because only natural kinds are real causal factors. In other words, Fodor presupposes that Putnam was partly right in arguing that reference depends on how the world is; but the problem is that to know what words refer to, we need to know everything about the real referents of all the words! And we cannot know that before we have completed all the sciences. For this reason, the theory of reference is practically impossible: we cannot wait for all the other sciences to develop if we want to pursue psychological research (Fodor 1980). But this is obviously a fallacious argument; we can and do talk of causal transactions without knowing everything about the causal factors involved.[1]

Another reason Fodor mentioned was that computational symbols are formal, which, on his construal, means that they cannot have content; so computational theories of cognition cannot deal with contentful states at all. However, as I pointed out in chapter 2, the formality of symbols in computational systems consists in the fact that they need not refer and not in the fact that they cannot refer. I hope the symbols I am typing right now refer to something other than machine instructions. Specifying exactly how that happens constitutes a challenge for a theory of representation that I want to take on in section 3.

Stated otherwise, Fodor's claim was that psychology cannot explain content and reference. Our best psychological theories are computational, so they are formal, so we cannot explain the reference of representation in these theories. I think this inference is valid: theories of computation are not in a position to explain reference. But it is not sound. Psychological theories are computational in a different sense than Fodor supposes in that they do not require formality. If a given physical process is computational, it has physical properties that are not formal. And the problem is that uninterpreted formal symbols hardly deserve the name of representation.

Newell's physical-symbol hypothesis is a fairly classical account of information processing in cognitive science, and it relies on a classical, von Neumann computer architecture. But if Newell means something similar to Fodor, his hypothesis does not pertain to any functions of representation, such as having reference, content, being correct or wrong, and so forth. The symbols remain uninterpreted. One might take it as a feature of this model of representation; hence the formalist motto, "if you take care of the syntax, the semantics will take care of itself" (Haugeland 1985, 106). Taking care is playing by the rules; the symbols and the rules of their manipulation are all there is to meaning.

John Searle's Chinese room thought experiment (Searle 1980) fervently attacked the formalist assumption. The Chinese room sparked a huge discussion, and different people use it in different ways; however, it needs be stressed here that the original story about the Chinese room was supposed to undermine research on artificial intelligence and computer simulation of human cognitive capacities. I'll quickly summarize the thought experiment and its significance to discussions concerning representation. Let us imagine John Searle, who does not speak any Chinese, closed in a room with a special set of rules that enable him to answer, in Chinese, questions posed in Chinese. The gist is that Searle is able to produce, without knowing the meaning, Chinese output in a completely formal way. There were many objections raised to the conclusions that Searle had drawn from his thought experiment. I will not dwell on them here, as they usually focus on the possibility of building artificial intelligence in general (see, e.g., Chalmers 1996b, chapter 9). Most important for my discussion is the conclusion that Searle draws for artificial systems: all the meaning there is in such systems is merely ascribed from the outside by external observers.

Searle's conclusion is based on the putative analogy between the human being and the computing system operating on formal symbols. The most popular reply to Searle, the Systems Reply, is that it is not the individual component part of the system, such as the processor or the human being operating on symbols, but the whole system that is the locus of understanding. To say otherwise is to commit a localization fallacy by ascribing a system-level capacity to a component part (see Bechtel and Richardson 1993 for a description of the fallacy in general). It does not seem to be obvious that component parts, as such, understand or refer to anything; and it is definitely not a conceptual truth about representation that it is to be localized in the individual component parts of some system. Otherwise, the only viable theory of meaning in neuroscience would be the theory of so-called grandmother cells, which refers to the individual cells that activate in the presence of referents of representations or complex stimuli—for example, in the presence of a grandmother. (For an early expression of this view in terms of "gnostic units," see Konorski 1967; for a historical review, see Gross 2002.) In other words, Searle needs to justify his claim that understanding is to be ascribed to component parts of the system and not to the system as a whole. Searle's answer to this objection is not satisfactory at all: he says that it is enough for the individual to internalize the Chinese room with all the rules and so forth. But in such a case, when the system is operating sufficiently fast, there seems to be no in-principle difference between an individual who uses symbols and a cognitive agent that has neurons that play exactly the same role (see Hofstadter and Dennett 1981, Chalmers 1996b).

It was also pointed out that the notion of "understanding" as used by Searle is notoriously vague. Does it mean that symbols need to be interpreted, or that they need to refer, or maybe that they need to be accessed in a conscious manner, which is what Searle seems to endorse (see also Dennett 1987, 335)? This, however, seems to beg the question. Searle simply assumes, without any further justification, that consciousness has to be biological and that no machine could have it.

Another problem with Searle's thought experiment is that it is merely a story, not an argument. In Dennett's words, it is just an intuition pump, and a misleading one, as the story is incoherent and relies on intuitions that are connected with size and the speed of processing (Hofstadter and Dennett 1981). In subsequent discussion, Searle (1992) reformulated his argument in more discursive terms by claiming that computer programs are purely syntactical and that syntax is never sufficient for semantics. For this reason, as he claims, no computer program can be said to mean anything, as it is only formal.

The notion of formality gives rise to a subtle equivocation that seems to underlie Fodor's methodological solipsism as well (Fodor 1980, see chapter 2, section 3). In one sense, symbols are formal as they need not refer, and in another, they cannot refer at all because they are formal. The latter claim is patently false for computer symbols; the former, however, is not enough to establish the conclusion that formal systems cannot refer (or be semantic). So the whole argument is fallacious.

But the fallacy notwithstanding, Searle's thought experiment seems to grab attention anyway (Dennett 1996). It is hardly believable that reference is just a matter of relations between symbols that need not refer at all. Such symbols might be ascribed some meaning, but only from the point of view of an external observer. The meaning might as well be completely epiphenomenal as far as the operation of the system is concerned, so it could not play any explanatory part; if it is not a causal factor, it cannot explain how the system works—at least not in the mechanistic account of explanation.

In subsequent discussion, the argument that symbols are meaningful simply because they are (proper kinds of) computer symbols became a

rarity. It would be as credible as saying that a Chinese monolingual diction-
ary explains what Chinese words mean to people who do not speak
Chinese. (This is not to say that conceptual [or rather functional] roles
cannot be informative about some aspects of meaning.) In cognitive
science, Searle's thought experiment was used to formulate a specific
problem—the symbol-grounding problem. "How can the semantic interpretation of a formal symbol system be made intrinsic to the system, rather
than just parasitic on the meanings in our heads?" (Harnad 1990) The very
existence of this question means that Newell's optimism about physical
symbol systems was misplaced. The systems were physical, and they contained the symbols, but they were not organized in a way that would make
it clear that the symbols possessed content. The existence of the mind in
the physical universe can hardly be explained in this way. For this reason,
Good Old-Fashioned Artificial Intelligence (GOFAI) (Haugeland 1985) was
also called into question. The claims of GOFAI, as Haugeland defines them,
are that (1) "our ability to deal with things intelligently is due to our capacity to think about them reasonably (including subconscious thinking)" and
(2) "our capacity to think about things reasonably amounts to a faculty
for internal 'automatic' symbol manipulation" (Haugeland 1985, 113).

Although it is widely believed that GOFAI was a dominant view in classical cognitive science, it was never the only view. For example, Newell
and Simon, when discussing Bartlett's idea that motor skills are crucial for
thinking, noted that motor skills seem to be nonsymbolic (which seems
to contradict their later all-embracing notion of the symbol), which "makes
them a poor model for a system where symbols are central" (Newell and
Simon 1972, 8). This might be interpreted as restricting the scope of symbolic models to higher-level cognition only.

The perceived failure of GOFAI to account for meaning led to various
attempts at solving the symbol-grounding problem (for a review, see
Taddeo and Floridi 2005), most of which started with the assumption that
there must be some link between the artificial system and its environment.
Obviously, many cognitive scientists rejected the notion of symbols altogether and began to work with other kinds of representations—for example,
the ones found in connectionist networks. But the background theory they
have of representation makes this problematic, and eliminativism or anti-
representationalism do not seem to be very exotic options (see, e.g., Ramsey
2007). A set of commonsensical assumptions that are shared makes the

problem simply intractable. I will argue for rejecting these assumptions in the next section.

2 Natural Meaning to the Rescue

If the main problem with the formal symbols is that they are formal, then the obvious suggestion would be to replace them with symbols—or any other content-bearers—that are not formal and do not require an observer to be meaningful. This is basically the strategy adopted by the so-called Robot Reply to Searle (1980). It leads to a proposal to use natural meaning relationships—relationships that obtain between smoke and fire, and between a photo of Fido and Fido the dog. There are various ways to carve such relationships, but in general they are either relations of resemblance or covariance.

Theories based on resemblance relationships rely on a certain similarity between the object and its representation. Most such accounts appeal to a structural resemblance relationship (e.g., Craik 1943), and many writers are attracted to the notion of isomorphism in explicating it (see, e.g., Palmer 1978; S-representation in Ramsey 2007). Nonetheless, if isomorphism is understood simply as a one-to-one mapping, then even a Fido's photo is not isomorphic to how Fido looks, since Fido is a 3D object, and his photo has only two dimensions. For this reason, homomorphism seems much better suited to account for structural resemblance (Bartels 2006), since a many-to-one mapping suffices for it to obtain. A full-blown theory should also focus on preserving the structure of the object being represented (or specifying the exact conditions needed for understanding the notion of morphism). Going forward, I shall assume, for the sake of argument, that the notion of resemblance is formally definable.

The second class of theories of representation focuses on the covariance of events, of which causation is used most frequently for this purpose. Just as smoke signifies a fire, all effects represent their causes. This relationship of content determination appears to comport with how people think of perception: that percepts are caused by things external to the mind. No wonder, then, that causation (or covariance in general) is usually thought to be instrumental to how the mind represents these objects. This can be linked with an account of information flow (Dretske 1982). It is also the classical account of cognition: representation is a sort of symbol that is the result of information processing occurring in the sense organs (Miller, Galanter, and Pribram 1967, 50). In other words, there are some transducers of sensory signals into symbols (Pylyshyn 1984). Neisser writes:

The term "cognition" refers to all the processes by which the sensory input is transformed, reduced, elaborated, stored, recovered, and used. It is concerned with these processes even when they operate in the absence of relevant stimulation, as in images and hallucinations. Such terms as *sensation, perception, imagery, retention, recall, problem-solving,* and *thinking,* among many others, refer to hypothetical stages or aspects of cognition. (Neisser 1967, 4)

It might be supposed that the notion of covariance, or a causal account of representation, is best suited for explicating the notion of representation used in cognitive science.

However, both covariance and resemblance accounts of representation are flawed for several reasons. Peirce noticed that icons (which are similarity-based representations) and indexes (covariance-based representations) are not exactly the same as other signs—they lack the interpretant, or content (for an accessible review of the Peircean taxonomy of signs, see Short 2007, chapter 8). A dyadic relationship between the referent and the sign cannot specify the referent of the sign. But as long as we understand "content" as what makes representation's referring possible, and do not conflate reference and content, it's not easy to know where or what the content is. I am not implying that explicating the notion of content is an easy task that only lazy theorists forget to do properly; I simply want to point out that dyadic accounts make this even more difficult. Moreover, resemblance is generally a symmetrical relation, but we usually don't think representation is symmetrical. If a picture is similar to Fido, then Fido is similar to the picture; therefore, Fido refers to a picture. Another point is that these relations are reflexive; the best representation of Fido would be Fido himself (given that identity implies similarity with every respect), and his covariance with himself is also highest (which is why causation, as antireflexive, is much better suited to explicate representation). One might stipulate that a representation must not stand for itself, but apart from being ad hoc, this would rule out all kinds of self-reference, including such innocuous cases as the word "word" representing all words. Transitivity is another problem: some similarity relations are transitive, some not; the same goes for covariance (and causation). In the simple case of isomorphism, the relation is necessarily transitive, for example. Yet representation in general need not be transitive; the expression "a picture of Fido" represents a picture of Fido, and a picture of Fido represents Fido, but "a picture of Fido" does not represent Fido.

A more general problem with natural-meaning relations is that they are conceived of as encoding (see Bickhard and Terveen 1995) or as a certain mapping (usually one-to-one). However, these approaches cannot account for how representation comes into being. For example, how should I know whether a picture represents Fido or some other dog? In order to answer this question, I would first have to know that the picture was intended to represent Fido. How else could I know that? It seems this account is viciously circular. We cannot tell what we need to consider when determining whether there is a mapping if we do not already know that there should be a mapping between two entities. But to know that there is a mapping, we would have to know these two entities; the notion of representation was supposed to tell us which entity the representation is related to. For this, we could use an external observer, but then we would have the same vicious circle of observer-based representations as with formal symbols. In other words, we need to know how to pick the target of encoding independently of the encoding itself (Cummins 1996).

Mappings abound in the world because the world is full of information. All regularities and laws, including laws of arithmetic, constitute natural mappings that such accounts would construe as representations (Ramsey 2007). The resulting notion of representation will always be too broad. Imagine that an evil scientist takes a picture of Fido and then opens my skull and inserts the picture into my brain. The picture is not *my* representation (not to mention that it is not mental either) even though it does represent Fido and is connected with my brain causally (unfortunately for me). But that causal connection is not enough.

Encoding-based accounts are also too narrow. Empty representations have no referents, so there is nothing that they can relate to. A picture of a unicorn cannot represent a unicorn by virtue of resembling a unicorn, because there *are* no unicorns, so nothing is similar to them. One could object that we develop such empty representations by way of putting together several representations that actually do refer (say, "horse," "white," "gold," "horn," and "one") and claim boldly that all empty representations have the component structure required for such an analysis. But the problem reappears as inability to misrepresent. A natural sign cannot misrepresent; if, for example, smoke never occurs without fire, then smoke as such cannot help but to refer to fire. Any failure to represent must therefore be an illusion produced by the sign's interpreter, who might misidentify something as a natural sign (say, by taking a cloud of dust for smoke). This, incidentally, constitutes a general challenge to all informational semantics (Godfrey-Smith 1989).

My summary of how natural meaning is used to naturalize meaning can be criticized in the following way: these positions are too simple; for example, it is possible to defend homomorphism-based theories of representation that allow for misrepresentation (Bartels 2006). We only need to distinguish reference from content by introducing the distinction between the target (referent) and all the other things to which the vehicle is structurally similar (for a proposal on how to do this, see Cummins 1996). That might be so, but then natural meaning is not the whole story about representation. The story about coding and decoding understood in simplistic causal terms is one of the most popular theories of meaning in both cognitive science and neuroscience, so I do not think this is a straw-man position (for similar remarks, see Ramsey 2007 and Bickhard and Terveen 1995).

I do not mean to say that natural meaning or intrinsic information in the world is irrelevant to the representational capacities of cognitive systems. Discovering how information is actually encoded and decoded in complex neural systems is certainly relevant to validating the information-processing view of cognition, but it is also extremely difficult due to our inability to measure the information flow directly (for an excellent review of state-of-the-art methodologies that make reasonable estimations and idealizations possible, see Rolls and Treves 2011). It is certainly explanatorily relevant to know whether representations are distributed in neural populations, which is, as Rolls and Treves argue, typical in primates, or sparsely encoded in the brain (as in the grandmother-cell theory). But establishing these neural facts will not suffice as a theory of representation, even if it can show that some kinds of representation cannot be stored if there is not enough structural or selective information content in the information bearer. Representation cannot be reduced to natural meaning construed as a kind of mapping relation (just like computation could not be understood in terms of a simple mapping only; recall the discussion in chapter 2, section 2). Moreover, the proper naturalization of representation will not only need to say something about content determination and misrepresentation but also about how representation is constituted in the first place. It must be clear not only what the content and referent of a representation are but also what its job is in the cognitive system.

3   Toward a Theory of Representational Mechanisms

Until now, I have been relying on our intuitive understanding of "representation" and using the concept freely. However, because the term gives rise to as many mutually incompatible interpretations as the word "symbol" does, I need to introduce a terminological distinction that will come in handy in subsequent discussion. I want to distinguish between information and representation. The difference should be somewhat familiar: a representation must represent something *to* the cognitive system—as a whole or to its part (called *consumer* in Millikan 1989; see also Bechtel 2001; Cao 2011 defends a teleosemantic account of representation that is closest to the one I offer in this section). In other words, it must be ready to be utilized *as a representation*. Contrast this with the picture of Fido shoved into my brain: I cannot utilize it as a representation of Fido in this setting even though it certainly plays a causal role. All natural-meaning relationships constitute intrinsic information that, if ready to be utilized in a special mechanism, might become a full-blooded representation.

Not surprisingly, I will spell out my conception of representation in mechanistic terms. I contend that representation cannot simply float freely without being part of a highly organized system. The role of the mechanistic theory of representational mechanisms in the philosophy of cognitive science is to enable us to analyze the notion of representation—to describe the organization of the hypothesized mental mechanisms whose job is to represent and evaluate such descriptions. So the theory is normative as well, as is the general account of mechanistic explanation itself (see Craver 2007). However, my intention is not to offer yet another philosophical analysis of the concept in question; what I provide is just a preliminary sketch, which I need in order to discuss the explanatory roles of representation in the computational theory of cognition. Some existing theories of representation might be treated as special cases of my approach; this is especially true of the ones that focus on the role of representation in behavior.

My sketch will be contentious. The mechanism I describe is a semantic engine. Some would say that this is a "mechanistic impossibility—like a perpetual motion machine, but a useful idealization in setting the specs for actual mechanisms" (Dennett 1991a, 119). But this semantic engine is not divorced from the syntactic engine, which is physically possible. Indeed, the claim that representation requires computation will be justified in the following paragraphs. It must be noted, however, that despite the link I suggest exists between syntax and semantics, semantic features are not epiphenomenal in my model. They are causally and explanatorily relevant.

The goal of the preceding discussion was not only to show the inadmissibility of popular theories of representation, but also to gain some insight into what a theory of representation is expected to deliver. Any mechanistic explanation starts with a specification of the mechanism's capacity, and we now have a fairly clear list of desiderata. Its main objective is to show the *function* of representation (and its content) in the cognitive system. To put the point differently, the mechanistic theory of representation should clarify the distinction between information and representation, and show how information may serve to represent. The account should also make misrepresentation possible; misrepresentation is still a kind of representation, from which information should be distinguished (note that misinformation is *not* a kind of information according to Dretske 1982). The cognitive, rather than purely functional, role of misrepresentation must also be made clear. In other words, we do not want to introduce yet another external observer to tell us whether or not a system misrepresents. Error should be system-detectable (Bickhard 1993; Bickhard and Terveen 1995).

The insight about error being system-detectable bears emphasis, for it suggests at least one role content can play in the cognitive system: if a piece of information serves as a representation in the cognitive system, then the cognitive system will care about its epistemic value because a representation is useful only insofar as it is reliable. This requirement narrows down the class of representations in the set of information having a causal role in the system. For example, if the representation is to guide action by predicting the state of the environment, the system should be able to detect a discrepancy between its predictions and the state of the environment. In other words, interactive representation (Bickhard and

Terveen 1995; Bickhard 2008), as well as action-oriented (Mandik 2005; Wheeler 2005) or action-based representation in the guidance theory (Anderson and Rosenberg 2008), specifies something along the lines of the representational mechanism in my sense; an early version of this idea is also to be found in the analysis of meaning in terms of readiness to react (MacKay 1969).[2] In general, there is also some convergence with other models that posit certain control structures to explicate representation (e.g., Grush 2003, 2004) or see perception as poised for action (e.g., Webb and Wessnitzer 2009).

The distinctive causal role of representation is salvaged if the system is itself sensitive to both misrepresentation and referential opacity. Intensionality is a mark of genuine causal explanations involving representation (Dennett 1969). Nevertheless, the discussion of causation might become a red herring in debates over representations. For example, William Ramsey (2007) presents a challenge for theories of representation: such theories ought to show what their particular role is (job description challenge). However, in evaluating the solutions proposed, he seems to adopt a sort of causal exclusion argument. In discussing Dretske's account, Ramsey assumes that if something has a causal role, it does not serve as information. A similar argument has been voiced by Andy Clark and Mike Wheeler (Wheeler 2005) and by Brian Cantwell Smith (2002). But it is trivially true that causation transfers information and that it is impossible to transfer information in any other way (if you believe in causality; see Collier 2010 on causation and information). If something plays a causal role, it is not at the same time excluded from being information. On the contrary, all causation involves information. But only some causation, and only some information, generates cognitive representation for the system; this is what the potential use of information is. The consumer of information is not just any physical system but a system that is sensitive to error in information.

"Causal spread" is a problem only for a purely causal account of representation, such as the one based on semantic transduction (section 2 above). As Cantwell Smith rightly observes, it requires that everything beyond the system be nonsemantic and that it become semantic on entering the system; otherwise, inexorable problems line up. But the problems stem only from the assumption that semantic properties are an outcome of some special causal relationships related to transducers. This is an assumption I reject. The world contains structural-information-content (MacKay 1969; see chapter 2) before it gets into the cognitive system; there is no need to deny that. But it becomes a cognitive representation only when playing a proper role in a representational mechanism.

What is a representational mechanism? What role does it play in the cognitive system and how does it work? A representational mechanism is capable of the following:

• Referring to the target of representation; that is, identifying the object that is being represented (if any)

• Identifying information about the object that is relevant for the cognitive system (i.e., content)

• Evaluating the epistemic value of the information based on the feedback from the environment

It is important to note that such capacities may be realized by various mechanisms, which are built according to a number of organization principles. The mechanism might include a discrete vehicle of information as its component part, for example, but the vehicle might just as well be a relational structure of intercomponent connections (i.e., a distributed vehicle). The content itself might be encoded in some vehicle of information, or it might be dynamically generated. A classical model of representation in cognitive science posited several features of representation:

(1) being enduring, (2) being discrete, (3) having compositional structure, (4) being abstract, and (5) being rule-governed. It is now known that representations may or may not have these properties (either singly or in various combinations) depending on where in the cognitive system they do their work. (Dietrich 2007, 11)

Note that my model of the representational mechanism does not decide whether representation has these features or not. This is an empirical question about the organization of various cognitive systems and should not be decided a priori.

One way in which the standard causal accounts deal with misrepresentation is by bringing in teleological considerations (see Dretske 1986). My model also stipulates that representation is functional, but I am relying on a different notion of function (Krohs 2007; see also chapter 2 for an explanation of how function is integrated with the notion of mechanism). This notion of function does not make the occurrence of functional properties a purely historical fact, which might be inert (i.e., preempted in causal explanations by proximal factors). (As Dennett 1991a argues, historical facts about meaning cannot be causally relevant; for a discussion of the epiphenomenalism of function in Millikan's account, see also Bickhard 2008.) However, malfunction and misrepresentation are two different things. Take thermoreception (Akins 1996). Human thermoreceptors, when functioning normally, are not veridical as indicators of the temperature of one's surroundings. They overreact, and with good reason—being extra careful with hot and icy things pays off. The same goes for any situation where false positives are less costly than false negatives. This consideration can be used to argue that truth and falsity are not the only epistemic values of representation. As a matter of fact, we do not always pursue the truth. When controlling a nuclear plant, for example, we should allow for numerous false alarms, which are useful in the overall context of the thoroughness of detection of possible nuclear leaks.

The system-relativity of epistemic goals is one of the reasons why representations are not simply copies of the environment. Though it is still quite popular to talk of representational function as supplying a stand-in for relation (see, e.g., Grush 2004), this is not what representation does according to my model. The first reason for this is that information structures are used differently from what they are about (Sloman 2010, 2011); for example, a recipe for a cake cannot replace, or stand in for, a cake, and a map is not a copy of a city. This difference between the information bearer and its referent is crucial for the utility of information in cognition. Second, if all representations had to carry all the information available in the world, then one might use the world just as easily instead. But the role of a representation is to reduce complexity and eliminate noise. For example, a map will not inform you about the cars in the satellite photo that was used to produce it; cars are (*ceteris paribus*) noise for a cartographer. Third, reducing complexity and selecting relevant pieces of information is in itself a cognitive achievement that justifies storing the information for further use. It is computationally cheaper to store information for reuse than it is to recompute it if the processing is sufficiently complex. However, this means that all accounts of representation in terms of strict isomorphisms might fail—representations will exclude some information about the world, and this will be their feature rather than a bug. The information content of a representation, therefore, is not at all about simply mirroring, copying, or preserving all there is in the world. This does not mean that preserving some relation structure of reality has nothing to do with representing; I do not deny that the function of modeling is precisely to preserve some structure (for an early argument, see Craik 1943, 51). But a theory of representation should focus on relation structures rather than on preservation (note that Craik [ibid.] talked about similarity instead of identity or isomorphism), and selection of information makes the relevant structures more salient.

It is also arguable that the epistemic value of representation varies relative to the goals of a cognitive system. If the cost of representational accuracy is high in that it involves, for example, lengthy and time-consuming cognitive tasks, and a quick and dirty approximation is enough for achieving a goal, then it is more rational for the system to prefer the quick and dirty approximation. As the goals of the system influence the evaluation of a representation, it is natural to connect representation with the behavior and action of the system.

This is not to say that representations are just causal mediators in motor behavior. This kind of account might be challenged by pointing out that not all representations affect motor behavior (Prinz 2000; Clark 2001b). Cognitive neuroscience provides obvious counterexamples to a purely motor account of representation. Milner and Goodale (1995), for example, argue for a deep disassociation between conscious vision and the visuomotor system. If there are two different systems—one responsible for motor behavior and the other for what we consciously experience as seeing—then it would follow that the latter is not representational according to motor accounts of representation.[3] This might be true of radical sensomotor conceptions but not of the model I am proposing. As Anderson and Rosenberg (2008) stress, the notion of action they appeal to is not necessarily an immediate motor action. It might be cognitive. One obvious objection to such a notion of action is that it conflates action with any kind of activity. I concede that this use of the notion of action might be mildly misleading, and this is why I have not included it in the specification of the representational mechanism. At the same time, if representations are functional and not epiphenomenal, then they must play a causal role in the activity of the cognitive system. In the long run, cognitive systems do act, though representations do not have to influence their actions directly. On the contrary, the possibility of decoupling behavior from stimuli is what motivates positing representations as explanatory factors.

The more a representation is decoupled from current environmental stimuli, the more explanatory burden it will take. Reacting to nonpresent conditions is most easily explained with representations. As Andy Clark says, off-line adaptive behavior is a mark of representational abilities (Clark 1997). However, it would be premature to think that the distinction between present and nonpresent is a demarcation line between the system's representational and nonrepresentational capacities. Obviously, there must be a causal link between past conditions and the present states of the cognitive system so that the off-line capability is not a result of a sharp discontinuity. It would be wrong to presuppose, along with antircp-resentationalists (e.g., Calvo 2008), that continuities in cognitive processes critically undermine the role of representations. They don't; it is rather decoupling between reaction and environmental conditions that is important. If there is no possibility for real error, as in the example of a photo-taxic behavior of a slime mold (Anderson and Rosenberg 2008), there is no real decoupling. It is the possibility of error that introduces the real need to posit representations, and the off-line adaptive behavior is just one corollary of the ability to make representational errors.

Note also that, unlike Bechtel (1998, 2001), Chemero (2000), or Nielsen (2010), I do not think that any control structure that includes information about the system under control will be representational. For example, the notorious Watt governor cannot really detect error in its information (nor can the system that contains just the Watt governor and the steam engine). A Watt governor (see figure 4.1) is a device that controls the speed of the steam engine, and its role is to stabilize the steam engine by opening or closing the valve:

The governor is so adjusted, that when the engine is working at its normal speed, the balls rotate at a certain distance from the vertical spindle, and thus the throttle valve is kept sufficiently open to maintain that speed. Should the load be *decreased,* the speed of the engine, and therefore that of the governor balls, naturally becomes greater. This causes an increase of the centrifugal force of the balls, and therefore they diverge further, thereby pulling down the sleeve and partially closing the throttle valve, which diminishes the supply of steam and the power developed by the engine. On the other hand, should the load be *increased* the reverse action takes place, the balls come closer together, the sleeve is raised, the throttle valve opened wider, and more steam admitted to the engine. It will thus be seen that a change of speed must take place before the governor begins to act; further, that for any permanent change in the work to be done, there is a permanent alteration of speed.

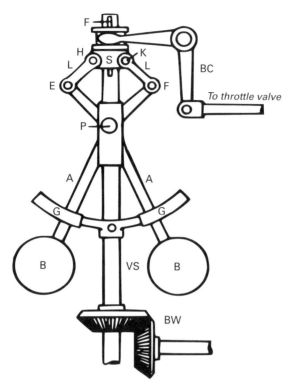

Figure 4.1

Watt Governor. "This governor consists of two arms, A A, carrying heavy balls, B B, and pivoted on a pin, P, passing through the centre of the vertical spindle, VS. The upper ends of these arms are bent, as shown on the figure, and are connected by short links, L L, to the sleeve, S. This sleeve is free to move vertically on the spindle, VS, but is made to rotate with it by a feather, P, and corresponding keyway. This sleeve acts on one end of the bell crank, BC, and thus moves the rod connected to the throttle valve of the engine. The vertical spindle may be driven by the engine by means of a belt or rope passing round a pulley keyed on it, or by bevel wheels, as shown at BW. In order to relieve the pin, P, the arms are driven by the guides, G G, which are fixed to the vertical spindle" (Jamieson 1910, 402–403).

For each particular load on the engine, the throttle valve will be opened by a definite amount, which will be different for different loads, and each position of the valve has a corresponding position of the governor balls. But, as will be shown further on, each position of the balls corresponds to a definite speed, so that there will be a particular speed for each different load. (Jamieson 1910, 403–404)

In modern language, we would say that to stabilize the speed, the governor needs reliable information about the current state of the engine; this is accomplished by connecting the spindle to it. (For the exact modern explanation of the informational content, carried by the balls, that causes the sleeve to go up or down, see Nielsen 2010.) There is, however, no additional feedback on the reliability of this information or on the success of the control action. There is negative feedback that helps the Watt governor to open or close the throttle valve; this is how this device works. However, it cannot detect that closing the valve did not help in speeding up the engine, as there is no additional source of information about the state of the valve. This would be required for the Watt governor to detect some kind of representational error. In other words, I do not deny that the Watt governor is an information-processing mechanism. However, I would not say that its complexity is sufficient to call it representational in the full-blown sense.

If you think of representationality as a graded property (see Bickhard 1998, who studies these grades as "levels"), then the Watt governor would be just minimally representational. Much like a thermostat, it could be described from the intentional stance (Dennett 1987). It is also similar to a thermostat in that it has negative feedback, so it can be said to have functional presuppositions (Bickhard 1993), or evaluatory mechanisms (MacKay 1969, 67), but a representation at this level cannot yet be in error *for* the system; both thermostats and Watt governors are organized in a way that does not allow them to detect errors in their states of readiness to react to information.

Not all goal-directed systems that use negative feedback (such as thermostats or Watt governors) have representational mechanisms. In contradistinction to standard thermostats, a genuinely representational system is capable of detecting that its action, as based on some previous state, had an unanticipated result. So if a thermostat discovers that the temperature in a room is not influenced according to its predictions and switches to a backup temperature sensor (we can imagine a special redundant system similar to the ones used in nuclear plants), it can be said to have discovered a possible error in its temperature readings and in its representation of future temperature. In general, systems that rely on forward models (Webb 2004) or predicate about the future state of the environment (Bickhard 1993) do contain appropriate evaluational mechanisms and, hence, are representational.

The significance of evaluation can also be appreciated from the teleose-mantic perspective. As Cao argues, the consumer of the representation, or the receiver of the information, is necessary for mental representation; the sender, or the source of information, is not sufficient by itself (Cao 2011). Moreover, to call something a receiver, one needs to find the adaptive function of the information it receives; there must be some reward in receiving the information:

> The receiver needs to have interests—not in the sense of being conscious or aware, but rather, in the sense of having states of affairs that matter to it, so that there are "better" and "worse" outcomes in every situation. A receiver has to be the kind of thing that can be rewarded . . . the receiver needs to be an agent, again, not necessarily in the sense of having intentions, but at least in the sense of being able to act in the world to affect its own outcomes. An incoming signal will only carry semantic information for its receiver if the receiver has the ability to act on the information in a consequential way, even if that ability is not exercised often, or even perhaps has never been exercised. (Cao 2011, 53–54)

These considerations lead Cao to stress that reward can be appreciated only by entities with interests, which is typical for living things. This is just another way of stating Bickhard's point that the role of representation is normative: error is important as far as it can be detected by the agent, and for that, the agent needs to have interests (Bickhard 1993).

In a sense, the mechanistic account of representation as meaningful *for* the system might be read as following Searle (1984, 1992) in his insistence on intrinsic intentionality. But in Searle's theory of intentionality it is utterly unclear why only conscious biological systems should be genuinely intentional. I wholeheartedly agree with Dennett here (1987, 334); as far as justification goes, Searle might just as well have contended that only left-handers' brains produce intentionality. My point, therefore, is not that representation needs to be intrinsic or genuine in Searle's sense. Further, I do not mean to advocate any kind of subjective view on representation involving first-person acquaintance with information or phenomenal con-sciousness.[4] No, my point is that the system should be organized in such a way as to both care about information in general and be able to detect error; whether or not a given system possesses these features is a perfectly objective matter of fact that can, by way of empirical inquiry, be established by scrutinizing the organization of the system and its behavior.

The condition that a system capable of representation be able to detect error can be directly related to the internalism/externalism debate in the philosophy of language. According to internalism, the content of a representation is determined within the cognitive system. Externalism denies that notion by pointing out that no cognitive system can have complete control over what it refers to; for example, try as might, I cannot make "Santa Claus" referential. Causation of rational behavior was the main reason why internalists (e.g., Fodor 1980) maintained that we need to focus on the internal dynamics of the system. It is hard to see why an action of someone who believes in Santa Claus could be explained by saying that the belief does not refer to anything. In other words, internalists wanted to point out that there is a certain opacity, or intensionality, to our thoughts, which we overlook when talking about reference in externalist terms. Fodor thought that only the intrinsic features of a cognitive system could be implicated in causal transactions; he maintained that only intrinsic properties are causally relevant, which is simply not true (see Wilson 2004 for multiple counterexamples from physics and biological sciences).

Let me rephrase the argument in simpler terms. If we want to explain why a child does something (e.g., tries to be polite) in the belief that otherwise Santa will not give her presents, we do not want to determine the content of her belief in a way that would make it indiscernible from my beliefs about unicorns. If she does not know that there is no Santa, then the emptiness of the term cannot be involved in the explanation of what motivated her to act. At the same time, it would be absurd to think that "Santa" really refers to Santa, and our talking about her beliefs always involves externally mediated language, which makes specifying the internal (narrow) content especially problematic. (For an explanation of the reason why the notion of narrow content is not useful in a nonencoding approach to representation, see Bickhard 1993, 323–325.) It is also clear that it is Santa's nonexistence that might explain why a polite child never gets any presents from Santa. The belief that Santa exists might therefore be a fallible guide to action: some actions will fail because representations fail to refer.

The discussion between externalism and internalism is usually spelled out in locational vocabulary. The two positions are primarily interested in *where* to look for content-determination relations. Both seem to focus on the input to the cognitive system, thereby losing sight of the temporal dynamics of cognition, which is precisely what gives rise to representation. In my account, the content of representations is information that is relevant for the system, and only this information is also causally relevant to explaining the action of the system (i.e., how the content is used to guide the activity). The facts about the referent of a representation, in general, may involve factors external to the system (self-referential representations might be less externally influenced). However, the information the system possesses can be decoupled from reality, giving rise to the possibility of the system being wrong about the world—wrong in a way detectable via failure of its subsequent activities. The representation itself is not reducible to either the source of information in the environment or the input to the information-processing system; its content's connection with the input might be very complex.

This is quite evident given the multiple levels involved in mechanistic explanation. When confronted with a specific representational mechanism, one has to account for its contextual level, which will inevitably include the environment of the mechanism, and that means the explanation will have to contradict the internalist or formalist principles. The role of the representational vehicle—and content—will be included at the isolated level. This would accommodate a valuable insight of internalism, namely that the explanation of many aspects of behavior has to focus on the kind of content that is available to the system. There is a certain asymmetry between the perspective of an external observer and the internal perspective of the system. Information within a representational mechanism can affect the system's behavior only if the system has access to it.

At the same time, the success of a system's activities depends not only on its representations but also on the way the world is, which is to say on the way the mechanism is embedded in the environment (described at the contextual level). This is why we cannot explain it by looking only at the content of a representation. To put the same point slightly differently, the internalists were right in their insistence that what an external observer knows about the referent of a representation cannot be causally relevant to the system's behavior if the system itself lacks such knowledge, but they erred in thinking that content could be completely decoupled from environmental factors—we can and do explain facts about perception by looking at how it is related to the environment. Even if thermoreception results in signals that are determined not only by the state of the environment but also by the previous state of thermoreceptors in the skin, we cannot simply skip the environment in our explanation.

Consider the frog—an animal which often features in philosophical discussions of signal detection (inspired by Lettvin et al. 1959; for a recent review of what frogs can represent, see Schulte 2012). Suppose the frog has a built-in bug detector which gets activated in the presence of small flying objects; upon activation, the detector initiates a cascade of events culminating in the frog catching the object with its tongue and eating it. Now, what would be the content of the frog's representation? Is it enough to know that the frog eats small buttons thrown in the air by an evil scientist to say that it has fallen victim to misrepresentation? In my account, it is not. As long as the frog itself is happy and swallows the button without ever realizing it is not a fly, there is no misrepresentation involved, even if we—external observers—know perfectly well that plastic buttons are not edible (assuming, of course, the frog has not been genetically modified to digest plastic). The event would be an instance of misrepresentation only if the content of the frog's "belief" were something like "This is an edible fly-looking object," which it can only be from *our* perspective and not the frog's.

Suppose now that the evil scientist throws a dead mosquito in the air, and the frog catches it. Suppose also that the mosquito is edible and nutritious, although it comes from outside of the frog's normal habitat. Would we still be tempted to say that the mechanism has misrepresented something? I don't think so. The mechanism does not appreciate certain differences in the world because they have no bearing on the frog's culinary goals. The bug detector may get activated in the presence of buttons because they do not occur in the frog's natural habitat, and other inedible flying objects are rare; therefore, it had been cheaper for evolution not to build a flying-button detector that would prevent the frog from eating pieces of plastic. My point, therefore, is that what counts as content depends on how the system can evaluate the representation guiding its behavior. This is the only nonepiphenomenal kind of content that might ever prove explanatorily relevant to the activity of the frog. But frogs, contra Schulte (2012) and other philosophers, might be too similar to thermostats to even qualify; their behaviors, just like the behaviors of crickets (see next section), appear to be easily accounted for in terms of taxis toward the stimulus (e.g., Guazzelli et al. 1998).

What lessons do I propose to draw from the frog *gedankenexperiment*? First, it is clear that a plausible explanation of how contents work in the frog must rely on information about the frog's discriminatory abilities on the one hand and on its capacity to peruse representations in guiding behavior on the other. Such an explanation will not appeal to human theories about plastic buttons because there is no causal chain stretching between humans and frogs that would enable us to transfer information from our scientific theories to frogs. Second, it is equally obvious that a completely closed physical system would have no need for representation of the external world. And cognitive systems need to be at least partially open, for otherwise they would not be cognitive. The causal pathways in the environment abound in all kinds of information (which can be cashed out in terms of the infomorphisms of Barwise and Seligman 1997; see Collier 2010), and a cognitive system can exploit that. Contra Fodor (1980) it is possible, then, to provide a causal account of the processes that allow a cognitive system to utilize external information. Note that a description of what the information is about will include distal causes, which are sometimes more explanatory than immediate proximal causes (McClamrock 1995). So, in some ways, you could say that representational mechanisms have an internalist flavor to them, even if the overall story is decidedly externalistic; both how the system is embedded in the environment and how it acts in that environment are important.

Since representation involves information, the operation of representational mechanisms requires information processing. Note that all the requirements I gave in the previous chapter for information processing are fulfilled: there is a functional mechanism, the mechanism operates on information (which has a physical vehicle), and a complete account of a given representational mechanism will also provide the details about the format in which the information is cast. A story about how a representation is produced on the basis of input information will be largely computational. There is no representation without computation. But the story neither starts nor ends with computation: although, arguably, the inputs might be parts of the environment (as adherents of wide computationalism, such as Wilson [2004], argue), the functioning of a representational mechanism requires more than computation. The proper functioning of the evaluation subsystem cannot be explained exclusively in computational terms.

Two capacities of representational mechanisms rely essentially on a causal relationship between the system and the environment: (1) The capacity to extract information about the target, which the mechanism then uses to refer to the target via content; and (2) the capacity to evaluate this information. Note that these are two distinct capacities—asserting that the same thing happens in both cases would amount to making the mistake of correspondence-as-copy theories. If representing consisted in making a copy and then evaluating it by making a second copy in order to check the one against the other, the whole process would yield no real additional value (see also Bickhard 1993 on skepticism). So, basically, extracting information about the target must be conceived in terms of feedback from the environment. However, the notions of causation, feedback, or control are not computational. The point might be subtle, because physical computers are causal and they include control structures; nonetheless, an explanation of why they are causal is not computational, but physical.

Note that if you understand computers as mechanisms that implement formal models—not to be confused with the formal models themselves, which is what I argued for in previous chapters—you can easily integrate the explanation of computation and the explanation of representation. There is no great gap between computation and representation, though arguably representational mechanisms are not equivalent to computers. They include computational capacities, but they also interact with the world, usually in three ways: (1) By extracting the information from the world, (2) by influencing the world via representation-driven activities of the system, and (3) by getting feedback from the results of such activities. (There might be more interaction involved, and the feedback might be also hierarchical, although engineering several orders of feedback is difficult.)

In many cases the notions of detection or recognition, employed so eagerly in cognitive science and neuroscience to account for representation, can be assimilated into the mechanistic framework. It is important to remember in this connection that any theory of representation the researchers happen to espouse should be distinguished from whatever genuine contributions they may have made in their field; reinterpretation constitutes a viable option here, especially if some elements of the representational mechanisms at issue are presupposed tacitly (for a similar argument, see Bechtel 2001).

I will briefly sketch how reinterpretation of explanatory models in my representational framework could proceed. The standard story about representation is told in terms of coding and decoding, where the coder is the mechanism responsible for extracting information from the environment. This is a confused way of speaking, because the information is not simply encoded and preserved, but rather detected and selected from noise. A more credible theory, therefore, might be framed in terms of the detectors of real patterns (Dennett 1991b). Being sensitive to information is the same as being sensitive to detectable differences. The same experimental methods that give insight into information extraction are usually employed to justify the claim that there is covariance or a causal relationship between sensory receptors and the features they respond to. (Note that a real pattern detector might be sensitive to highly structured information; standard examples in introductory literature focus on individual bits of information to simplify the discussion.) If this is where the story ends, the account remains incomplete and the extracted information can hardly be said to be representational for the system.

If, however, we have evidence suggesting that the system exploits the information to accomplish some further tasks (as in visuomotor processing), and we also have a hypothesis about its role in guiding the behavior, then the information becomes representational in character. It is also important that the theory provides for correcting the information by way of feedback from the outcome of a behavior. This part is usually left implicit. However, the role of feedback and error detection was acknowledged as early as in MacKay (1951) and Miller, Galanter, and Pribram (1967, 51). I am not really inventing the wheel here. I am just pointing to parts of the story about representation that were simply forgotten or ignored in the classical computational theory of the mind.

4   Explaining Representation and Explaining with Representation

In cognitive science, representation has two explanatory roles to play (Ramsey 2007). One is the role of *explanandum* phenomenon; the representational mechanism must then be explained in terms of the organization of its parts and their causal interactions. The other is the role of *explanans*. In such a case, it is shown, according to the mechanistic model of explanation, how the explanans contributes to the explanandum phenomenon. The contribution is spelled out in organizational and causal terms. The very representational phenomenon, however, is not the focus of the explanation, and it might remain quite sketchy. Note that the explanations of cognitive tasks focus on tasks as *explananda*. Representations figure therefore mainly as *explanantia*. For this reason, it is to be expected that representation will be taken for granted in such theories of cognition. In what follows, I will go through the cases I introduced in chapter 1 and show various uses of representation in explanations.

Cryptarithmetic tasks, such as SEND + MORE = MONEY, are explained by Newell and Simon (1972) by appeal to a computer model that is based partly on behavioral data. Their analysis of the task relies on problem spaces, which are specific, crucially representational posits. A problem space consists of a symbol structure and the operators that transform it. Solving the task is construed in terms of searching this space. There are three problem spaces that Newell and Simon examine: basic, augmented, and algebraic space. The basic space contains (1) letters and digits that are used to build expressions, (2) knowledge states that are built out of expressions, (3) a set of knowledge states U, (4) an operator Add, and (5) a set of operators Q (ibid., 145). The operator Add simply adds to a problem space a hypothesized assignment of numerical value to a letter of the alphabet. There are further conditional operators on the problem space that may yield a solution; the details need not interest us here. The rough idea is that one can test whether the solution has been found in the space by testing first if the assignment of all digits is complete and whether it produces a valid arithmetic operation on digits after translation.

The augmented problem space is not limited to storing specific assignments; it can contain such additional information as constraints on assignments (like $R > 5$, or that it is odd or even). In other words, the total set of states of knowledge is larger. In the algebraic problem space, the problem is represented immediately as a set of algebraic equations. The condition DONALD + GERALD = ROBERT is equivalent to the following list:

$$2D = T + 10c_2 \qquad\qquad (1)$$

$$c_2 + 2L = R + 10c_3 \qquad\qquad (2)$$

$$c3 + 2A = E + 10c4 \tag{3}$$

$$c4 + N + R = B + 10c5 \tag{4}$$

$$c5 + O + E = O + 10c6 \tag{5}$$

$$c6 + D + G = R \tag{6}$$

The search of the solution will therefore be equivalent to solving this set of equations. Newell and Simon stress that these three spaces are most frequent among the subjects, but the list is not complete and it would be hard to provide a formal demonstration of its completeness (Newell and Simon 1972, 156). After analyzing the available spaces a priori, they proceed to the analysis, based on a protocol, of human data on a single subject. Interestingly, they distinguish between an internal and external problem space:

> The subject writes some things down, but long periods go by with no writing. Thus, there is an internal problem space that is clearly different from the space of written, externalized actions. (Ibid., 167)

They also describe the problem space based on the behavior of the subject—this is the augmented problem space, which is used by most technically educated people—and build a production system based on the protocol. What is interesting from the current point of view is that the external and symbol spaces are investigated on a par; in other words, a proponent of the conception of extended mind would rejoice at that. The explanation is in no way limited to the spatial boundaries of the individual. This is also clear in the pioneering use of eye-tracking methods.

It is quite clear that problem spaces are explanantia. Newell and Simon explain task solving rather than problem spaces, so this was to be expected. The problem itself is rather representation-hungry: it requires sensitivity to highly abstract properties (for more on representation-hungry problems, see Clark and Toribio 1994, especially pages 419–420), and a classical symbolic rule-based explanation is offered. Note also that although assignment of numbers to letters might be construed as a discovery of what human subjects designate by symbols (letters), Newell and Simon do not specify the task in this way. Maybe they saw the idea that "D" really designates 5 as too far-fetched.

In the second example, the task is learning the past tense forms of English verbs (Rumelhart and McClelland 1986). The artificial neural network solves this problem by starting with a specific representation of the input data consisting of structurally rich phonological patterns—Wickelphones. Wickelphones are based on a formal analysis (Wickelgren 1969), so in this respect they resemble problem spaces posited by Newell and Simon (1972). In learning, only individual verbs are used, which makes the task as artificial as cryptarithmetic, in that it does not really correspond to real-life grammar learning. The output representation is also couched in terms of Wickelphones, which means that the network matches the scheme of explanation as conversion of input information into output information. Interestingly, the features that the network learns are not straightforward representations of external objects—they refer to the phonological structures of phonological information. There is no commonsensical concept that would correspond to such features, and one can easily see why they could be called "subsymbolic representations." They are explanantia, as this model does not explain why these representations actually underlie our acquisition of the past-tense forms. However, one might tacitly suppose that certain feedback mechanisms are posited—it seems obvious that, at least in speech perception, we realize that some of the words were not perceived correctly. Whether the connectionist network can account for such top-down influence of the mental lexicon is another question, but, in principle, it should.

All in all, the representation used here simply relies on what Wickelgren (1969) suggested. For this reason, some claim that the operation of the model is trivialized. The objection is that "the performance of connectionist models is dependent upon particular ways of encoding inputs which are borrowed from other theoretical, usually symbolic theories" (Bechtel and Abrahamsen 2002, 126). As Bechtel and Abrahamsen say in their reply, much processing remains to be done once an encoding scheme is decided upon. Moreover, Wickelphones, though constituting a smart solution to the encoding problem, are not biologically realistic. However, they have some advantages from the mechanistic point of view when they are encoded as Wickelfeatures: "Phenomena at one level (e.g., acquisition of past-tense morphology) are best understood in terms of mechanistic models at a lower level (here, phonological features)" (ibid.). It should also be noted that the model can be regarded as disproving the claim that symbolic rules and representations are necessary for past-tense acquisition, and that Rumelhart and McClelland's work is to be interpreted not as an explanation but as a proof of possibility. (For a similar reading of this work, see also Bermúdez 2010, 259.)

The third example is much more interesting as the representation acts both as explanans and as explanandum. The rat navigation model (Conklin and Eliasmith 2005) relies on the Neural Engineering Framework (NEF), which is committed to representations. The NEF takes "the central problem facing neuroscientists to be one of explaining how neurobiological systems represent the world, and how they use those representations, via transformations, to guide behavior" (Eliasmith and Anderson 2003, 5). As before, representation is accounted for in terms of encoding and decoding: "Neurons encode physical properties into population-temporal neural activities that can be decoded to retrieve the relevant information" (ibid., 8). This may sound like a commitment to encodingism, and indeed, the NEF provides a specification of what encoding and decoding consist of (see chapter 1, section 4). However, this is not the end of the story: "Knowing what is represented depends in part on how it is subsequently used, it seems like we already have to know how the system works in order to know what it represents" (ibid.). This is precisely what my model of representational mechanisms requires. In other words, the NEF uses the engineering notions of encoding and decoding to talk about information. Representation, however, is not being reduced to encoding, whereas decoding is actually understood both as part of neural representation and as subsequent transformation of the neural representation. The important fact is that neural representations are control-theoretic state variables.

The current estimated location of the rat is modeled as a two-dimensional bell-shaped function that corresponds to a bump of neural activity. The neural representation is part of the control mechanism, and the overall model includes a high-level mechanism responsible for navigation. While exact mathematical details are to be found in the original paper, the most interesting question from my point of view is whether the neural representation can be in error *for* the rat. The answer is in the positive: error correction based on visual cues (when available) is considered as possible feedback from the hippocampus (Conklin and Eliasmith 2005, 189, 196). It is also notable that the researchers distinguish between mathematical representation (such as projections into the Fourier space), which are convenient abstractions of the state of the neural network that permit the application of control theory, and the real neural network. They stress the biological plausibility of the vector representations they use (ibid., 191).

Moreover, one of the important results of modeling is that the control mechanism encodes not only the position of the rat on the plane but also a two-dimensional in-plane velocity vector. This leads to predictions of the existence of velocity-sensitive place cells in specified locations of the rat's brain.

In other words, the neural representation is used here not only to explain how the rat navigates, but also to explain it as such. Detailed hypotheses about distributed representation are empirically well-founded and modeled by means of a novel attractor-network model of path integration (this uses heterogeneous spiking neurons, which is also much more biologically plausible than early connectionist neurons). In spite of the vocabulary of the theory, which is reminiscent of encoding-based assumptions of representation, representation is related to controlling the behavior and is methodologically plausible according to my account of representational mechanisms. Note also that even though computational explanation is a crucial part of the model, it also requires—at the contextual level—describing cues available to the rat and positing certain causal relationships that allow for extracting information from these cues. This is precisely what is impossible according to the formality principle of Fodor's methodological solipsism.

The fourth example is of a robotic model of cricket phonotaxis (Webb 2008). The model reflects the morphology of the real cricket's ears and is physically well adapted to detecting the frequency of the song of cricket males. The wavelength of the calling song (6–7 cm) compared to the separation of the ears (1–2 cm) makes locating the sound problematic. But the cricket's ears contain a solution: a pressure-difference receiver; therefore, the physical makeup of the model allows for delaying and filtering the sound. In other words, the input information is modeled in a biologically plausible way by replicating the essential properties of the morphology. Webb and Scutt's (2000) neural model is geared toward specific neural circuitry of the cricket (four neurons are used) and is at the level of membrane potential (spiking neurons are used in the simulator). The output of the circuit that processes the auditory input is fed into the motor circuit. The robot is able to replicate a number of real crickets' behaviors, which validates the model.

Now, the question is whether the auditory signal in the robot is representational. Note that the information is not stored (crickets do not seem to remember the location of a past sound, they simply follow the current sound), which means that the stimulus is present throughout the whole taxis. Although input information plays a control role in the robot, there do not seem to be any feedback mechanisms, which is why it is more like a Watt governor in that it is processing information but has no representation that could be in error. There is not enough decoupling between the signal and the motor response for that, although there is some evidence that invertebrates use forward models; these models account for some predictions being made by the nervous system about the future states of the motor system based on current sensory input and previous motor states (Webb 2004). The prediction can be confronted with real input from the motor system; that is the basis for the detectability of errors in the system. It may turn out that cricket models will have a use for real representations; at this stage the robot model is very simplified. And, as I mentioned in chapter 1, given this lack of representation, one may indeed deny that robotic phonotaxis constitutes a cognitive phenomenon.

5 Conclusion

In this chapter, I sketched the role of representation in computational explanations. There is computation without representation, but there is no representation without computation. Information processing is a prerequisite of representing, and yet information is not the same thing as representation. To analyze the uses of representation-like notions and posits in cognitive science I hypothesized several general requirements for all representational mechanisms. One of the essential features of these mechanisms is that they are able to detect error in a representation. Another is that they are embedded in the control structures of cognitive systems.

In some ways, my account of representational mechanisms flows naturally from the analysis of the drawbacks of earlier popular models. There is a role for computation—inferences and other operations on information are an important dimension of representational mechanisms. But such information remains purely formal. It has to be connected to the world. However, this connection cannot be made by simply adding natural information from the world to noninterpreted symbols. Natural signs as such are not full-featured representations, as they cannot be in error. A mixture of external information and formal symbols would not suffice either, though such "two-factor theories" remain attractive for philosophers

(Block 1987). I submit that such theories might tell us something about the determination of content but have no real role for it in the cognitive system. For this, the account has to encompass the behavior of the cognitive system and the epistemic value of representation for the system.

In traditional cognitive science, cognition was equated with information processing over representations. As I argued above, representational mechanisms involve computational processes, so information processing over representations is what is always involved when representations are used in a cognitive system. However, only a robust notion of a representation as guiding the action of the cognitive system by supplying it with proper anticipations—the notion that I vindicated in this chapter—may define the mark of the cognitive. So even if I agree that representation suffices for cognition, which is a traditional stance indeed, I mean something different from the views held by the mainstream theorists of cognitive science. In my version, representational mechanisms are sufficient for cognition; this implies that more than computation is required for cognition, for such mechanisms cannot be reduced to purely computational mechanisms. At the same time, I think that, for theorists who rely on representation (or similar concepts) to define the mark of the cognitive, this is the notion of representation that would serve better than the notions of information processing or nonderived content (Rowlands 2009; Adams 2010).

I do not claim that representation is necessary for cognitive processes to obtain. Although nonrepresentationalists usually make their case by deflating cognition (Kirchhoff 2011), it is still an open question whether we want to call purely reactive processing, such as that present in the robot that imitates phonotaxis, cognitive. The decision to include reactive processing—or minimal cognition in bacteria (van Duijn, Keijzer, and Franken 2006)—under the umbrella of cognitive research may well be a terminological or programmatic one. I do not share the enthusiasm for defining "cognition" once and for all. Wimsatt (2007) and Machery (2011) argue that biology does not need a definition of life. Machery points out that, taken as a folk concept, "life" has no definition because folk concepts in general are indefinable, and providing a scientific definition of it would be pointless, for there are a number of disciplines that are unlikely to converge on a single vision of what constitutes the essence of life. Indeed, that is what makes them different disciplines! For Wimsatt, a scientific realist, this is just a corollary of the fact that "life" refers to a natural kind, and we are more likely to discover the essential properties of natural kinds than fix them with theoretical definitions. These considerations also apply to the definition of cognition: yes, we need to inquire into the essential properties of cognition as seen in various theories and disciplines that investigate it; but we are not likely to settle on a single, all-encompassing definition of cognition for all disciplines in cognitive science. Such a definition would be feasible if the disciplines in question were theoretically unified; alas, in this interfield enterprise called "cognitive science," it is more than likely that in the years to come disintegration will be as strong as integration.

5 Limits of Computational Explanation

The purpose of this chapter is to discuss several limits of computational explanation. For example, it is impossible to computationally explain the physical makeup of a cognitive system or completely account for its actual performance of cognitive tasks. The performance of a computational mechanism depends not only on the actual algorithm realized, even if it is specified in a fine-grained fashion, but also on the physical properties of the mechanism—a number of which cannot be understood in purely computational terms. Certain properties of the environment, or cognitive niche, may render some algorithms more feasible than others; these properties also escape computational explanation.

To answer the question about the limits of computational explanation, one needs to know what this explanation involves. I am going to show in this chapter that, in my mechanistic account, only one level of the mechanism—the so-called isolated level—is explained in computational terms. The rest of the mechanism is not computational; indeed, according to the norms of this kind of explanation, it *cannot* be computational through and through. Although numerous objections to computational theories of cognition will prove correct, the fact that this is so will by no means undermine these theories; instead, such objections will lead to a version of explanatory pluralism. This is especially true of representational mechanisms, which I introduced in chapter 4.

It will also be instructive to see how the four theories we have been discussing throughout the book fare when confronted with criticism—especially regarding their programmatic assumptions. Theoretical scrutiny has raised a number of serious doubts regarding the classical models of Newell and Simon as well as the early connectionist models of Rumelhart and McClelland. And models such as the Neural Engineering Framework or the biorobotics proposed by Barbara Webb are not immune to criticism either. After reviewing some objections that can be easily accommodated by the mechanistic framework, I will address, though only briefly, several problems that have been raised to challenge the very idea of cognitive science and AI. I argue that most of them are merely red herrings. At the same time, by drawing on examples from radical embodied cognitive science, I show that, contrary to popular wisdom, some purportedly non-computational mechanisms do feature computational processes. I do not treat these mechanisms as exclusively computational, however; they are multidimensional, which is what any descriptively correct theory of cognitive science needs to acknowledge.

1   The Scope of Computational Explanations

Constitutive mechanistic explanations of computational systems require at least three levels of organization; the levels under discussion are not mere abstractions or descriptions—they are levels of composition (see Craver 2007). The uppermost level (also known as the contextual level or +1 level) is the level of the mechanism's interaction with its environment. The explanandum phenomenon usually appears under specific environmental conditions, and the capacity of the mechanism plays its role only as long as these conditions are satisfied. The capacity of the mechanism is also described in terms of the isolated level, or 0 level. This is where the workings of the mechanism are investigated without recourse to its surrounding context. At the constitutive (or –1) level, the mechanism's components and their interactions are themselves analyzed to understand how the mechanism is constituted.

Note that, from the explanatory point of view, the contextual level of a computer is not computational. A description of how a computer "behaves" in its environment is couched in generic causal terms and not in terms of information processing. Of course, one computational mechanism can be a component of a greater computational system, in which case the former is described at the constitutive level of the latter, while the latter serves as the contextual level for the former. Integrating such computational mechanisms may consist in specifying both of them in computational terms; however, the uppermost level of the larger mechanism, as well as the constitutive level of the submechanism, will still remain non-computational in character.

By contrast, the isolated level of a mental mechanism is computational: one can specify it solely in terms of information processing unless one also wants to understand how it interacts with the environment. Moreover, the organization, activities, and interactions of the components of a computational structure, as it is represented by a mechanistically adequate model of a given computation, are also described in computational terms. As I emphasized, the description is correct only when all the components of a mechanistically adequate model have their counterparts in a generic causal model of experimental data. An explanation will be regarded as complete only if the objects and interactions at the constitutive level can be "bottomed out," or explained in noncomputational terms. For example, the electronic mechanisms of a calculator may explain how a certain model of computation is actually implemented: a mechanistically adequate model of computation is an explanandum phenomenon which must be explained as a capacity of the electronic mechanism of the calculator. So there is some bottom level that is no longer computational.

This will also be true of highly complex hierarchies: a computational submechanism might be embedded in a larger mechanism and this larger one in another. A submechanism contributes to an explanation of the behavior of a larger mechanism, and the explanation might be cast in terms of a computation, but the nesting of computers eventually bottoms out in noncomputational mechanisms. Obviously, pancomputationalists, who claim that all physical reality is computational, would immediately deny the latter claim. However, the bottoming-out principle of mechanistic explanation does not render pancomputationalism false a priori. It simply says that a phenomenon has to be explained as constituted by some other phenomenon than itself. For a pancomputationalist, this means that there must be a distinction between lower-level, or basic, computations and the higher level ones. Should pancomputationalism be unable to mark this distinction, it will be explanatorily vacuous.

It is clear, then, that everything that can be computationally explained is situated at the isolated level. The constitutive level is computational in the sense that a mechanistically adequate model of computation describes the organization of its entities—but their operation is not explained in this way. Entities at this level serve as explanantia rather than explananda: they account for the computational capacity of the mechanism in terms of how it is constituted. In other words, the constitutive and contextual levels need not be computational, and they are not *explained* computationally. What lies within the scope of computational explanation is only the isolated level—or the level of the whole computational mechanism as such.

This is easily seen in the cases I referred to in this book. Newell and Simon (1972, 800) presuppose that an information-processing system has certain capacities, such as the ability to scan and recognize letters and numbers from the external memory while solving a cryptarithmetic task. These capacities may be realized, as they stress, by parallel perceptual processes rather than by serial symbolic processes requiring attentional resources. Similarly, the physical morphology of cricket ears, the frequency of the sounds crickets make, the way they walk toward the sound, and so forth are not accounted for by the artificial neural network in a robotic model (Webb 2008). In other words, the physical implementation of a computational system—and its interaction with the environment—lies outside the scope of computational explanation. It is the physical implementation that explains why there is computation in the first place.

The upshot for the computational theory of mind is that there is more to cognition than computation. An implementation of a computation is required for the explanation to be complete; explaining the implementation requires an understanding of how a computational system is embodied physically and embedded in its environment. (This is admitted also in traditional cognitive psychology.) More specifically, reaction time is only partly explained computationally; the complexity of the cognitive algorithm, as analyzed mathematically, cannot be used to predict the run-time efficiency of the algorithm without knowledge of the appropriate empirical details of the underlying hardware. Moreover, for relatively short input sizes, measurement error can make it impossible for us to decide empirically, on the basis of reaction time alone, which algorithm is being implemented.

Resource limitations are also impossible to explain computationally. Instead, they act as empirical constraints on theory; for example, Newell and Simon (1972) impose the condition that the capacity of short-term memory not exceed the limit of 7 plus or minus 2 meaningful chunks. To put the same point somewhat differently, in order to understand a cognitive computation and to have a theory of it, one needs to know the limits of the underlying information-processing system.

Moreover, all of the environmental conditions that influence a computation, such as feedback from the environment, are at the contextual level. It follows from this that representational mechanisms are not fully explained in a computational fashion; some of their parts are generic and are simply interactions with the environment. Even Fodor (1980) acknowledged that an account of reference to the external world must contain some model of the environment. What he denied was that such a theory, which he called naturalistic psychology, could be built.

It is important to note that these limitations do not undermine cognitive research. A discovery that complex information-processing systems can be explained in a very simple manner, which ignores most of their complexity, would be much more surprising. If we want to take complexity into account, the causal structure of the system will come to the fore and the computational facet of the system will be just one of many. But the cogency of the computational theory of mind does not rest on a bet that only the formal properties of a computation are relevant to explaining the mind. Rather, it is based on the assumption that cognitive systems peruse information, and that assumption is not undermined by the realization that the way these systems operate goes beyond information processing. This is a natural corollary in the mechanistic framework I adopted, and the prospects of computationalism have not become any the worse for that.

2 Are These Really Real Explanations?

The four cases I analyzed throughout this book are hardly uncontroversial. In particular, the two classical studies had come in for harsh criticism. Let us therefore look at the most common objections to these studies. We do this in order to make sure my analysis was not based on highly contested assumptions and to see what my theory of computational explanation might say about the possible flaws inherent in these cases.

The heuristic account of human reasoning was fervently criticized by Dreyfus (1972), who made it the model case of computer simulation in psychology. One of his objections is that Newell and Simon's programs lack generality:

The available evidence has necessarily been restricted to those most favorable cases where the subject can to some extent articulate his information-processing protocols (game playing and the solution of simple problems) to the exclusion of pattern recognition and the acquisition and use of natural language. Moreover, even in these restricted areas the machine trace can only match the performance of one individual, and only after ad hoc adjustments. And finally, even the match is only partial. (Dreyfus 1972, 82)

Interestingly, Newell and Simon (1972) explicitly admit that theirs is not a theory of pattern recognition and so forth. Furthermore, given the lack of a deep unifying theory of all mental processes, it can hardly be an objection that any particular psychological theory fails to cover all the mental phenomena one would like to understand. The focus of the theory of human problem solving was human problem solving, not pattern recognition; this restricted scope of the theory is justified by appeal to the observation that this is where sequential thinking can be modeled. The authors also hypothesize that motor and sensory functions are nonsymbolic and parallel (Newell and Simon 1972, 8, 89; Simon 1979b, 4–5).

The discrepancy between the verbal protocol and the machine productions is hardly a vice—especially when the match is around 80 percent. This is definitely better than chance, and eye movement data provide an even better match. It seems quite clear that the cognitive load of "thinking aloud" is higher than that of simply thinking, so omission of some steps in the protocol may result from limited cognitive resources (*pace* Dreyfus, who sees these gaps as an insurmountable problem). As to the requirement of a perfect match between the model and reality, if the scientific community ever chose to observe it, all modeling would immediately become unscientific.

Newell and Simon are neither unaware that microtheories have a very restricted scope (indeed, this is why they are called microtheories), nor do they see it as an advantage (Simon 1979b, xi). Lack of generality, more or less, comes with the territory of experimental psychology. They strive to create a general framework that could constitute a basis for unification (Newell [1990] took a step in that direction). Dreyfus's objection is partly right, then: the scope was limited, especially in the early work; however, the predicament of Newell and Simon's microtheories was typical of all empirical psychology. Contrary to Dreyfus, modeling a particular system is not unscientific. Recent methodological considerations point out that modeling single participants rather than averaging across subjects might be a wise strategy (Busemeyer and Diederich 2010; Lewandowsky and Farrell 2011, chapter 3).

Generally speaking, Dreyfus opposes the methodology of modeling human behavior by means of rule-based systems and thinks that heuristic programs (composed of production rules, such as General Problem Solver, GPS) will collapse under their own weight (Dreyfus 1972, 8). To borrow Mark Twain's famous phrase, the reports of the death of production systems have been greatly exaggerated. They are still in use today, and complexity is not a problem for them. However, the significance of what Dreyfus calls fringe consciousness (which is equivalent to what Searle [1992] calls Background, and Polanyi [1958] called tacit knowledge) for problem solving might be greater than Newell and Simon had presupposed. Dreyfus seems to assimilate problem solving into pattern recognition (somewhat along Gestalt lines), which seems to comport with Simon's later research agenda of finding meaningful chunks; chunking, or recognizing the meaningful ways to analyze problems, is the main mechanism for learning in the symbolic paradigm (see papers collected in Simon 1979b; a concise summary of the idea of meaningful chunking is to be found in Simon 1996, chapter 3). In particular, Simon (1996, 71–72) hypothesized that experts see the problem differently, which means that they represent it in a different way than novices. The expert's representation carves the problem at its joints by identifying its meaningful components, which are also manageable cognitively; achieving expertise is then accounted for in terms of learning how to chunk problems as experts do. In Dreyfus's terms, experts learn to see the pattern.

Also, the question of whether all thinking actually consists of sequential searches in the problem space is hard to answer; even Newell and Simon seem to deny that it is so (by excluding perception, etc.). It is possible that only some sequential problem solving can be modeled using their production-rule systems.

Newell and Simon have also been criticized by those who accept the symbolic paradigm for discrediting parallel processing in thinking in general; other production systems, such as ACT* and ACT-R, support parallel processing (cf. discussion of seriality in problem-solving theories in Boden 1988, 166–167). Simon thought of seriality as a feature of human attention, yet he admitted that there is no sufficient evidence for this claim.

Marr (1982, 347–348) voiced another objection to Newell and Simon's research agenda. Himself interested in what Simon (1979b) called immediate processors (i.e., largely automatic information processing that operates at a low level), Marr thought that mental arithmetic occurred at too high a level to lend itself to modeling. His assessment seemed motivated by considerations similar to those Fodor later brought up in connection with the alleged limitations of cognitive science: that what admits of computational explanation is only modular processes occurring in the input systems. This is in contrast to the central processes of nondemonstrative inference that make the human mind so special and so utterly mysterious (since we appear to have gained precious little insight into how central processes work) (Fodor 1975, 1983, 2000).

In his early work, Fodor (1983) justified this by appeal to the holistic (or isotropic and Quinean) character of central processes. Now, given Fodor and Lepore's (1993) critique of most holistic approaches, it might seem striking that Fodor's pessimism about the prospects for a computational theory of central processes persists. Indeed, the reasons underlying his diagnosis have changed. Instead of holding that semantics in general supervenes on syntax and endorsing methodological solipsism (Fodor 1980), Fodor now believes that context-dependence makes such supervenience claims false. It does, but then again, there had been no solid grounds for thinking that computationalism is in any way committed to internalism and semantics-over-syntax supervenience (McClamrock 1995; Wilson 2004). As I was at pains to show in chapter 4, semantics cannot be reduced to computation (hence there can be no purely computational theory of meaning). It does not follow from this, however, that there cannot be any serious cognitive science of thinking or problem solving or that all theorizing has to be limited to immediate processors or "cognitively impenetrable" modules.

To see this, one needs to remember that Newell and Simon (1972) stressed the role of the task environment: it is the task environment that constrains the degrees of freedom in human behavior. In other words, it is the limits to human rationality—including the limits imposed by the task environment—that make psychology possible. As Newell and Simon say:

It is this fundamental characteristic of the human IPS that accounts for the elusiveness of psychological laws. Man is the mirror of the universe in which he lives, and all that he knows shapes his psychology, not just in superficial ways but almost indefinitely and subject only to a few basic constraints. This view should not be equated with the extreme form of cultural relativism sometimes espoused in anthropology, or the "man can be anything he wants" slogans preached by inspirational psychology. The universe that man mirrors is more than his culture. It includes also a lawful physical universe, and a biological one—of which man's own body is not the least important part. The view simply places the constraints upon the possible forms of human behavior in the environment rather than attributing these constraints directly to psychological mechanisms. It proposes limits on knowledge and how to obtain it, rather than limits on the ability to perform according to knowledge that has been assimilated. (Newell and Simon 1972, 866)

In normal situations, as Newell and Simon stress, biological limits—which are the sources of both commonality and individuality in human subjects— are not exceeded, so the more relevant set of constraints comes from the task environment. Indeed, if we were to happily ignore the environment, which is what Fodor recommended in the 1980s, the variation and flexibility of human behavior would be almost infinite, thereby preventing any sort of prediction—computational or not. However, Newell and Simon's replicable results and the recent psychological research on habits (Wood and Neal 2007) strongly indicate the importance of the environment in predicting behavior. The environment, along with biological constraints, makes cognitive science possible. According to the norms of mechanistic explanation, ignoring the contextual level of mechanisms would amount to making their capacities meaningless, for it is only at the contextual level that we can learn why the mechanisms operate the way they do. Moreover, their work—by limiting the degrees of freedom or flexibility in experimental setups—might be made predictable in this way. As Boden observes, modularity theorists who oppose Simon and Newell, "who say that there can be no scientific account of cognitively penetrable phenomena have too limited a view of what counts as a scientific explanation" (Boden 1988, 172).

If one were to deny the role of rule-based explanations of problem solving, one could perhaps accept the claims of connectionism. So let me now look at how the model of Rumelhart and McClelland (1986) has been received. The past-tense debate is still in progress (see Pinker and Ullman 2002a,b; McClelland 2002; McClelland and Patterson 2002), though it started long ago with a seminal paper by Pinker and Prince (1988). Their criticisms can be summarized thus:

We analyze both the linguistic and the developmental assumptions of the model in detail and discover that (1) it cannot represent certain words, (2) it cannot learn many rules, (3) it can learn rules found in no human language, (4) it cannot explain morphological and phonological regularities, (5) it cannot explain the differences between irregular and regular forms, (6) it fails at its assigned task of mastering the past tense of English, (7) it gives an incorrect explanation for two developmental phenomena: stages of overregularization of irregular forms such as *bringed*, and the appearance of doubly-marked forms such as *ated*, and (8) it gives accounts of two others (infrequent overregularization of verbs ending in t/d, and the order of acquisition of different irregular subclasses) that are indistinguishable from those of rule-based theories. (Pinker and Prince 1988, 73–74)

Pinker and Prince show that there are different regularities in morphology and phonology, so a unified representation of verbs in terms of Wickelfeatures will not do justice to the existing patterns in language learning, which is not reducible to phonology only. Moreover, Wickelfeatures are structured but not sequential representations. It is only thanks to the output-decoding network that they are brought together to form an output sequence:

In fact, the output Wickelfeatures virtually never define a word exactly, and so there is no clear sense in which one knows which word the output Wickelfeatures are defining. In many cases, Rumelhart and McClelland are only interested in assessing how likely the model seems to be to output a given target word, such as the correct past tense form for a given stem; in that case they can peer into the model, count the number of desired Wickelfeatures that are successfully activated and vice-versa, and calculate the goodness of the match. However, this does not reveal which phonemes, or which words, the model would actually output. . . . The Wickelfeature structure is not some kind of approximation that can easily be sharpened and refined; it is categorically the wrong kind of thing for the jobs assigned to it. (Ibid., 93–101)

More importantly, the Rumelhart and McClelland model cannot handle the elementary problem of homophony:

If Rumelhart and McClelland are right, there can be no homophony between regular and irregular verbs or between items in distinct irregular classes, because words are nothing but phone-sequences, and irregular forms are tied directly to these sequences. This basic empirical claim is transparently false. Within the strong class itself, there is a contrast between *ring* (past: *rang*) and wring (past: *wrung*) which are only orthographically distinct. (Ibid., 110)

In short, Pinker and Prince undermined the empirical accuracy of Rumelhart and McClelland's model. At the same time, they admitted that the model was unrivaled in terms of the granularity of its predictions. However, since then the original model has been revamped. A hidden layer has been introduced, and Wickelfeatures have been abandoned (Bechtel and Abrahamsen 2002, chapter 5). Note that even Pinker has proposed a hybrid "words and rules" theory that includes both a rule-based component and a connectionist model of word lexicon, though it has never been implemented as a complete simulation, which means it cannot be tested as thoroughly as purely connectionist models (Shultz 2003, 97). However, for my purposes, these further developments are of minor importance. What is crucial is that the model lends itself to thorough testing and criticism. The lesson to be learned from the past-tense debate is that representation of data, along with its encoding and decoding, is critical. It has to be plausible, if not immediately amenable to experimental evaluation. Otherwise, a model that relies on a certain form of representation will fall short of being explanatory: it will be just an instance of (not really efficient) machine learning. The problem for connectionists, who adopt the perspective of parallel distributed processing instead of classical architectures, is that the important phenomenon of U-shaped learning may have resulted from a bias in the training data itself (Shultz 2003, 95–96)—the frequencies of verbs were skewed in the training data sets (see chapter 1, section 5, for details).

From the mechanistic point of view, the main drawback of Rumelhart and McClelland's model is that it provided no explicit method for testing the representation format. This is why it is akin to the sufficiency analysis proposed by Newell and Simon, although they had been careful to analyze the possible representations and to test their hypotheses about them against verbal protocols. Rumelhart and McClelland's model was also susceptible to the criticism that the task had been essentially simplified by introducing Wickelphones to represent the ordering of phonemes (see also chapter 4, section 4).

To sum up, both classical cases are clear examples of how-plausibly explanations. They fall short of being complete mechanistic models of the phenomena, although, quite surprisingly, the symbolic model seems to have been more motivated by empirical considerations, even if it did not offer any plausible (not to mention empirically justified) neural implementation. Let us therefore look at the more recent examples of explanatory models.

The explanatory model of path integration in rats (Conklin and Eliasmith 2005) is a clear example of how the Neural Engineering Framework (NEF) can be used. In contradistinction to classical models, it is neurologically motivated and its corresponding neurological properties are measurable. Although the neurological evidence about rats is vast, it is still far from complete, and some hypotheses of the model cannot yet be tested. For all I know, however, although speculative, they seem well-founded in neuroscientific knowledge.

Yet one could ask whether the mechanistic norms of explanation do not require that the proper model be robotic. There are biorobotic explanations of rats' navigational capacities (Burgess et al. 1997; Burgess, Donnett, and O'Keefe 1998; Burgess et al. 2000). As Burgess et al. claim, "the use of a robot ensures the realism of the assumed sensory inputs and enables true evaluation of navigational ability" (Burgess et al. 1997, 1535).

Aren't realistic models better? Yes. But is this robotic model more realistic than the one based on the NEF? The answer is a blunt "no." First of all, Burgess's robotic models are based on a computational model that fares worse than continuous attractor networks (Samsonovich and McNaughton 1997, 5917). Second, the claim about the realism of robotic models is actually weakly justified. As Barbara Webb remarks:

Robots are not less abstract models just because they are physically implemented— a two-wheeled robot is a simpler model of motor control than a six-legged simulation. What does distinguish abstraction in biorobotics from simulations is that it usually occurs by leaving out details, substitution, or simplifying the representation, rather than by *idealising* the objects or functions to be performed. (Webb 2001, 1047)

Indeed, there is minimal representation of biological details in the architecture of the robot used by Burgess et al. Nonetheless, there is some truth to the realism claim: building complete models, both robotic and computational, requires filling out the details that could have been omitted from purely verbal theorizing. More importantly, however, computational models are not restricted by the limits of current engineering. The materials available for building sensors, for example, do not constrain the details that can be included in a computer simulation. For this reason, computational simulations may be more fine-grained than robotic ones. To sum up, the explanatory models developed in the NEF are methodologically correct from the mechanistic point of view, and they may even give more insight where robotic engineering is less developed.

This brings me to the biorobotic modeling of crickets. If anything, Barbara Webb's model of cricket phonotaxis (Webb 2008) seems to be exactly what the mechanistic point of view would approve of. One might, however, object that the model is not really cognitive; phonotaxis is a very low-level capacity. Robotic models of human cognition (for a review, see Webb 2001) are scarce. Most models focus on simple capacities of simple animals. The question of whether the robotic methods will "scale up" to human behavior remains open.

Another point is that, in contrast to research on imaginary creatures, or animats, which dominates cognitive robotics, Webb does not model a complete animal but only a subset of its capacities (see Webb 2009 for criticism of animats; for a reply see Beer and Williams 2009). Animats are supposed to embody the general principles of cognition or species-independent invariants. Because many evolutionary solutions are shared across species, animat research seems to have sound methodological credentials, and the level of generality of such models might be greater than a microtheory of cricket phonotaxis. The downside is that, as Webb (2001, 2009) stresses, it is much harder to validate such general models, since no one has a very good idea of how to make them empirically testable. Right now the evaluation of animat models appears to rely mostly on the researchers' intuition about what should count as a good benchmark of adaptive behavior.

Let me summarize. Obviously, one can find some flaws in cognitive models, be it insufficient empirical evidence and testing, design flaws (like the choice of representation in the connectionist models), or low generality. In addition to the advantages of having strict and fairly complete descriptions of the phenomena, computational modeling of cognition is not without its perils (Farrell and Lewandowsky 2010). During the implementation of a theory as a computational model, the developer has to make a number of decisions. These decisions might be ad hoc solutions that are hard to dissociate from the original theory, so additional care must be taken to document them. Moreover, some models might be artificially simple due to some artificial regularity in the data that the model processes (this is the case of the past-tense acquisition learning in Rumelhart and McClelland's original model). Other models might be too complex to be understood easily, so they do not really have epistemic advantages for their users; this is Bonini's paradox (see chapter 3, section 4).

Most explanatory models of cognition are far from complete. They show what organization would be sufficient for producing a cognitive behavior (how-possibly models) and sometimes cite some neural or behavioral evidence (how-plausibly mechanisms). In most cases, they fall short of being how-really models, but that is no reason to reject them. Modeling—just like representation in general—involves loss of information, for only through abstracting away from some properties can it become illuminating. A pragmatic approach to modeling—like the one presented by Webb (2001), who recommends evaluation of biorobotic models on seven different, relatively independent dimensions—is therefore inevitable. Webb cites the following dimensions:

- Biological relevance
- Realism level
- Generality
- Abstraction
- Structural accuracy
- Behavioral match
- Medium

The evaluation of cognitive models is a nontrivial exercise, and a model that excels along all the dimensions might be hard to come by. This is why we need a systematic way to relate different models of the same or similar cognitive capacity. Webb also admits that integrativeness of models (which seems compatible with the unification considerations I cited in chapter 3) is important—exactly for this reason. We do need a way to pursue cognitive research pluralistically but without dividing cognitive science into a number of disconnected subdisciplines. Thus, explanatory pluralism will be vindicated only insofar as there are chances to unify the various mechanisms and models.

3   Red Herrings, Wild Speculations, and Watt Governors

Impossibility arguments about artificial intelligence and computers are notoriously bad. For example, Weizenbaum (1976, 271) claimed that, given the difficulty of the problem, it would be too costly, and thereby unethical, to fund research on speech recognition. Nowadays, we have speech-recognition software on smartphones and on the web; it may not be perfect, but it works for many purposes. Similarly, Bar-Hillel (1964) contended that it is impossible for machine translation to tackle some ambiguities of human languages. For example:

Little John was looking for his toy box. Finally he found it. The box was in the pen. John was very happy.

I pasted this text into Google Translate and got a decent, though grammatically imperfect, translation into Polish. The meanings were almost fine.[1] Once there were people who bet that, contrary to some of Herbert Simon's enthusiastic speculations, computers wouldn't defeat a chess master or prove an important theorem (now think of the four-color theorem). And, obviously, beating humans at Jeopardy, which clearly involves common sense, is out of the question. However, AI systems are capable of all this and more, so we ought to be more careful: if there is no mathematical proof that something cannot be done, any verdicts are mere speculation.

3.1   The Frame Problem

The frame problem is probably the most common objection to any computational explanation of cognition. However, what exactly the frame problem amounts to has become difficult to pin down. In its original formulation, it concerned the question of how to represent the effects of actions in logic without having to represent everything that does not change. This problem has been more or less solved in nonmonotonic logic (for book-length accounts, see Reiter 2001; Shanahan 1997). Besides, the issue was only relevant for logistic, representation-based AI, but not necessarily relevant for other trends in AI—not to mention cognitive science.

In its broader sense, the frame problem is the question of how to decide what is relevant for action. Now, although *this* problem has not been solved in logic (what does a general algorithm for computing relevance look like?), it is not quite clear, really, why it should vex AI or computational cognitive science (Shanahan and Baars 2005). It is hard to see why it should plague *all* computational systems. In the four cases I discussed, nothing like the frame problem or the general problem of relevance ever appears to arise; even the classical production system is not bogged down with an infinite number of inferences before it takes any action. This should be impossible according to some philosophers; production systems are implementations of the idea that thinking is based on heuristics and, by the same token, so is choosing the appropriate course of action. But as Wheeler suggests,

Are relevancy heuristics alone really a cure for the frame problem? It seems not. The processing mechanisms concerned would still face the problem of accessing just those relevancy heuristics that are relevant in the current context. So how does the system decide which of its stored heuristics are relevant? Another, higher-order set of heuristics would seem to be required. But then exactly the same problem seems to reemerge at that processing level, demanding further heuristics, and so on . . . It is not merely that some sort of combinatorial explosion or infinite regress beckons here. (Wheeler 2005, 180)

The real production systems do not contain an infinite number of production rules; even the slowest linear search is guaranteed to finish in linear time (for n rules, the pessimistic estimate would be to search for n rules). Alas, linear complexity is one of the lowest orders of computational complexity! To create a real combinatorial explosion, one would have to construct a mechanism that generates heuristics on the fly during the search, for then it could always create another heuristic before finishing the search for a previous heuristic. The only motivation that I know of to build such a mechanism would be to generate something like the frame problem in the system. Such a combinatorial explosion is not an inevitable feature of production systems at all.

It seems that the philosophers' frame problem is thought to ensue because of the unbounded flexibility of human action and reasoning. However, both infinite flexibility and unbounded rationality are received philosophical views on rationality rather than confirmed empirical results. There is a plethora of results suggesting that human reasoning is bounded and context-dependent (for an introduction to the idea of bounded rationality, see Simon 1996). More to the point, a cognitive architecture that would get swamped in thinking about what to do before ever actually doing it would be quite exceptional; a real system like that, built for the explanatory purposes of cognitive science, must be hard to come by. Moreover, we have reason to believe that the architecture of consciousness does not allow the frame problem to appear in the first place (Shanahan and Baars 2005).

3.2 Holism

It has been argued for a long time by people like Fodor (1983) that holism is a major problem for the computational theory of mind. Holism is supposed to be a hallmark of central cognitive processes of nondemonstrative inference; because of it, something like the frame problem was taken to be inevitable. Although Fodor was soon to change his mind about holism (Fodor and Lepore 1993), he still thinks that a theory of central, nonmodular cognitive processes is not forthcoming. I have already dealt with this line of argument in section 2 of this chapter.

There is, however, another argument from a version of holism criticized by Fodor and Lepore (1993); this holism is a corollary of the inferential theory of meaning, according to which the meaning of a proposition is the set of its implications. Now, obviously, this is not the kind of holism that is actually used in cognitive science and computational linguistics to describe meaning. Even thoroughly holistic theories of meaning usually assert that semantic relatedness comes in degrees—that there is a measure of relatedness of expressions (expressed as a real number; for a detailed description of how it is used in WordNet research, see Piasecki, Szpakowicz, and Broda 2009, chapter 3). In other words, this version of holism is mainly a philosophical straw man; real holism, however, is not a problem for cognitive science at all. Fodor's basic argument against holism goes like this: holism makes communication between people who have different beliefs impossible, because different things follow from different beliefs (so even if you have just one belief that differs from mine, your set of beliefs will mean something else). This is cogent only if you define meaning as a set of inferences (and there is no reason to do so, anyway; see chapter 4, section 1). But meaning relationships might be more complex than inferences in classical logic, which do not admit of degrees. And the very introduction of degrees makes this argument a nonstarter.

3.3 Dynamic Systems

It is sometimes claimed that dynamic explanations of cognition are distinctly different from computational ones (van Gelder 1995; Beer 2000). It is, of course, a truism that physical computers are dynamic systems; all physical entities can be described as dynamic systems that unfold in time. What proponents of dynamicism claim, however, is something stronger:

A typical dynamical model is expressed as a set of differential or difference equations that describe how the system's state changes over time. Here, the explanatory focus is on the structure of the space of possible trajectories and the internal and external forces that shape the particular trajectory that unfolds over time, rather than on the physical nature of the underlying mechanisms that instantiate this dynamics. (Beer 2000, 96)

It seems, therefore, that dynamic explanation will abstract away from mechanisms in general and from computational mechanisms in particular. And, indeed, it has been argued that dynamicism involves a covering-law type of explanation (Walmsley 2008). Yet the paradigm examples of dynamical explanations are also quite easy to interpret as mechanistic (Zednik 2011), and when they are not mechanistic, they fail to be genuinely explanatory (Kaplan and Craver 2011). More importantly, many proponents of dynamical explanations require that they also refer to localizable and measurable component parts of the systems under investigation (Eliasmith 2010; Grush 2004). It seems therefore not at all justified to say that dynamical explanations do not refer to the physical nature of the underlying mechanisms. Some do, some do not; I would not be so sure whether it is fine when they do not.

What is important for my purposes is that there is no conflict between dynamicism and my version of computationalism. If these are distinct explanations, there is no problem: they will usually focus on different aspects of cognitive functioning. Dynamical systems, in contradistinction to connectionist models, are not usually proposed simply as replacements for classical explanations, so there is no explicit conflict involved. As most real-life explanations do not pertain to the same explanandum phenomenon (the cognitive capacity itself is differently conceived), they may still be integrated in the overall picture of cognition.

I do not think that computationalism and dynamicism are mutually exclusive. As Newell (1973) stressed many years ago, psychologists try to play the twenty-questions game with nature and win; they think that you can discover the essence of cognition by simply building a list of dichotomies. You cannot, and integrating various explanatory models is more valuable than overplaying the meaning of methodological differences. Playing Watt governors against production systems seems silly to me. Note also that, by my standards, the properly explanatory models (for all intents and purposes, a Watt governor is not a model of any cognitive capacity whatsoever) that are cited repeatedly by dynamicists, such as Elman's modeling of language in a temporal manner (Elman 1990), are computational. After all, Elman used a connectionist model, and, though he found an interesting way to interpret its structure, it remained a piece of computational machinery. The same goes for many synthetic models, such as Beer's (2000) model, offered in animat research.

Dynamical modeling is often thought to preclude a traditional computational explanation. For instance, Thelen et al. (2001) explain a phenomenon seen in seven- to twelve-month-old infants. In Piaget's classic "A-not-B error," infants who have successfully uncovered a toy at location "A" continue to reach to that location even after they watch the toy being hidden in a nearby location "B." Thelen et al. question the traditional supposition that the error is indicative of the infants' concepts of objects or other static mental structures. Instead, they demonstrate that the A-not-B error could be understood in terms of the dynamics of the ordinary processes of goal-directed actions: looking, planning, reaching, and remembering. A formal dynamic model based on cognitive embodiment both simulates the known A-not-B effects and offers novel predictions that match new experimental results. This seems like a real challenge to traditional computational explanation.

But what is the strategy of Thelen et al.? They show that the phenomenon can be accounted for in terms of motor planning. This is a deflationary strategy. We need not appeal to higher-level mental structures at all, which suggests that Piaget's logical attempt to understand the phenomenon might have been misplaced. However, two comments are in order. First, motor planning does not preclude the processing of information. On the contrary, these researchers use the notion of information to talk about neural findings relevant to an understanding of action planning. The only way their approach differs from an appeal to traditional information processing is that their model is framed in dynamical language, and plans are considered not as discrete but as continuous, graded representations evolving in time. Second, my mechanistic approach does not favor explanations of behavior in terms of high-level cognitive activity over accounts that appeal to factors of motor activity—especially if the latter bring more explanatory value, parsimony, simplicity, and so forth. Therefore, even radical dynamical models, if they still uphold the claim that the phenomenon is cognitive (and not just, say, physiological), explain it computationally by referring to the dynamics of information processing.

What about cases where information plays no essential role whatsoever? There, I contend, the phenomenon under explanation is not cognitive. If there is a dynamical theory that conforms to all explanatory norms (including those of theoretical parsimony, generality, etc.), then such a deflationary strategy is welcome, though, in effect, it blocks computational explanation. The claim I defend here is not that all behavior is to be explained computationally; I have taken pains to show that there are limits to the useful application of computational notions. In other words, the existence of a dynamical explanation of, say, a Watt governor's control over the speed of an engine does not prove that computational explanation is always redundant. There is an important difference between the role that the Rumelhart and McClelland model played against the rule-based accounts; in the latter case, it was held that it is impossible to explain past-tense acquisition without appeal to high-level symbolic rules of morphology. But the claim was limited to this phenomenon only, and not to any behavior; this is why the counterexample was so important in this discussion (hence Pinker and Prince's efforts to show that it fails to explain the phenomenon). Yet dynamical explanations of noncognitive processes are not counterexamples to the general claim that cognitive processes can be explained computationally. So an attempt to generalize the line of argument in Thelen et al. in this way is clearly mistaken. Similarly, from the fact that some intelligent behavior does not need representational mechanisms, it does not follow that representational mechanisms are nowhere to be found (for a similar argument, see Kirchhoff 2011).

In this context, another feature of dynamical systems deserves mention. Dynamicists stress that it is easy to model real-time dynamics, and they suggest that computational models are "temporally austere" (Wheeler 2005); Turing machines are formal structures that do not seem to model time-related concerns in any other way than by introducing the discrete steps of a computation (Bickhard and Terveen 1995). This objection is based on the conflation of the physical implementation of a computational process, which is spatiotemporal, with the formal model of computation. Mathematical models of computation do not need model real-time dynamics if their purpose is to prove theorems in computability theory! But engineers who build machines need to take timing constraints into account; and there are areas of real-time interactions, even in classical computer science, where one needs to describe the time constraints in a more fine-grained fashion (e.g., think of the computer program that controls a complex nuclear power plant in real time). There is absolutely no reason to ban dynamical description from our toolkit. Control theory seems to be an especially good tool to model such problems.

3.4 Nonsequential, Cyclic Interaction

Sometimes it is claimed that computationalism portrays cognition as sequentially processing input into output, whereas real cognitive behavior is cyclical and involves interaction with the environment (e.g., Cisek 1999). Indeed, for expository purposes it is easier to think of computation as a sequence of state transitions. However, that is all there is to it—information processing in complex systems usually is cyclical.

3.5 Nonstandard Computation

Doesn't computationalism exclude quantum computation, analog computation, or hypercomputation? Or connectionist networks? Not in my account. I endorse transparent computationalism (see chapter 2).

3.6 Lucas and Penrose

One of the standard arguments against computational cognitive science and AI is based on the arguments of Lucas (1961) and Penrose (1989). In Lucas's words, "Gödel's theorem seems to me to prove that Mechanism is false, that is, that minds cannot be explained as machines" (Lucas 1961, 112). His reasoning goes roughly like this: Gödel's theorem shows that there is an unprovable but true sentence in a sufficiently strong logical system (one containing at least arithmetic). The machine cannot see that the sentence is true (it would have to prove it, but it cannot, as it is equivalent to the logical system that contains the sentence); the mind can. Therefore, minds can do more than machines. The problem with this argument is that it requires that human minds be consistent. Yet it cannot be proved that they are, and we have no reason to think so (Putnam 1960). So the argument is unsound. It was also proved that all possible variants of the argument must lead to a vicious circle, inconsistency, or unsoundness (Krajewski 2007).

3.7 Common Sense

One of the obvious problems for early AI programs was the modeling of common sense, and Hubert Dreyfus submitted that this was their core problem (Dreyfus 1972). This might be true of AI. Similar criticisms of computational modeling—that it makes reasoning too brittle and inflexible—were put forward by Ulrich Neisser (1963). In fact, one of the reasons connectionism was endorsed by so many people (including Dreyfus) is that it promised to offer graceful degradation instead of complete breakdowns and flexibility instead of the brittleness of symbolic models. True, we are far from being able to model common sense, but it seems that there are no a priori grounds to think it is impossible; contrary to Dreyfus, I do not think that information-processing modeling will always lead to brittleness. It is quite difficult to discover the features of the architecture of the mind by virtue of which the mind is able to reason in a commonsensical manner, but progress has been made in this direction since Dreyfus voiced his critique in 1972. After all, computers are now able to win on the game show *Jeopardy*.

3.8 Autonomy and Autopoiesis

Another objection to computational accounts of cognition posits that the cognitive agent has to be autonomous. One version of this argument goes as follows. For the system to be truly cognitive, it needs to be autonomous, because only autonomous systems have real functionality that is based on their normative features (Bickhard 1993; Bickhard and Terveen 1995; Barandiaran and Moreno 2008). Autonomy, as Bickhard and Moreno argue, requires that a complex system be far from its thermodynamical equilibrium; otherwise, it will dissipate and lose its structuring. So what is required for cognition is some kind of self-maintenance or metabolism. In other words, there is no cognition without (some features of) life. (The theory of autopoiesis does not appeal directly to thermodynamic or energetic properties; see Maturana and Varela 1980, 89.)

Is metabolism really necessary? Well, if it is, then the computational theory of mind will be incomplete; computers won't cognize without metabolism. But what about a metabolic source of energy for a computer or a robot? This is not a mere thought experiment; some researchers (Montebelli et al. 2010) actually suggest that microbial fuel for robots makes a difference:

MFC technology has the capacity to produce bioelectricity from virtually any unrefined renewable biomass (e.g. wastewater sludge, ripe fruit, flies, green plants) using bacteria. This provides robots with a degree of energy autonomy concerning choice of (non-battery) "energy recharging" resource. . . . Given the present state of the art, a critical limitation of this robot, motored by a biological-mechatronic symbiotic metabolism, is energy requirement. Individual robots are required to wait long intervals between bursts of motor activity. Many minutes may be required for relatively little movement. (Montebelli et al. 2010, 750)

A robot seeking "energy recharging" resources will be linked to its environment, so the proponents of 4E methodology should applaud this. One thing that makes me less than enthusiastic, however, is that Montebelli's robot's energy supply has precious little to do with the way the robot functions, which is to say that no clue is given as to how *cognitive* autonomy would emerge. If anything, this robot looks like a caricature of autonomy to me.

But autonomy, just like creativity (Boden 2004), remains an interesting topic and something of a holy grail of cognitive research. For my money, betting on energy is a bad investment. I have already stressed that nontrivial information-processing mechanisms are highly complex and nonaggregative. By my lights, these complex information-processing systems seem a better locus of cognitive autonomy; they need not be grounded in energy because functionality does not require thermodynamic autonomy (chapter 2). But the building of an autonomous and creative synthetic entity is still beyond the reach of current technology, and it is not clear whether or not it is possible within a purely computational framework. This is an open question.

4   A Plea for Pluralism

Using a single model of computation to explain all possible cognitive systems would be premature at best. Some cognitive systems, such as sponges or plants, may be explained in terms of simpler computation models, whereas more complex processes require interlevel explanations to give meaningful, idealized explanations and predictions. In other words, my explanatory pluralism involves both the claim that computation is not the only way to explain cognitive systems and the thesis that various models of computation might be useful in cognitive science, as it seems plausible that different models may best describe organization at the bottom level of the mechanism in various systems.

Interestingly, some have interpreted one of my four paradigm cases, namely Webb's research on cricket phonotaxis, as an example of a noncomputational model (Wheeler 2005). The only reason to do so was that the robot's morphology played such an important part. I do not want to play the same trick by interpreting the Webb model as exclusively computational—the artificial model, after all, relied on artificial neural networks, which did something important. Similarly, Elman's models of natural language (Elman 1990, 1991) rely on artificial neural networks. Granted, these are not traditional models of language processing, but innovative does not mean noncomputational. In similar vein, I classified Conklin and Eliasmith's (2005) model of path integration as computational, though it is arguably not a traditional exercise that posits some kind of language-of-thought cognitive maps.

The exclusion of some dimensions from a description of research should be motivated by something other than mere enthusiasm for other flavors of modeling. The decision to exclude information in models should be fruitful for our understanding of the phenomena, as all idealization in science should be. Otherwise, exclusion is, inadvertently or not, a kind of censorship. I tried to show in this book that successful examples of explanation are already pluralistic; they involve explanations of computation, physical structures, environments, and real-time interaction. I think this pluralistic character is one of the features of cognitive science as such.

An attempt to replace all explanatory methodologies with a single one seems premature. Progress in research is fueled by the cooperation between—and competition among—modelers using various modeling techniques. The previous accounts of the explanatory utility of cognitive models could not, however, accommodate this richness of methods. Traditional machine functionalism implied that implementation details are no longer interesting for researchers of cognition; although Marr (1982) stressed that implementation is a part of the proper explanation of computation, he did not rely on such details to explain anything. But over time, the traditional functionalist model has become less and less credible as the role of neural detail has been acknowledged. This does not undermine a kind of broad functionalism:

Neurochemistry matters because—and only because—we have discovered that the many different neuromodulators and other chemical messengers that diffuse through the brain have functional roles that make important differences. What those molecules do turns out to be important to the computational roles played by the neurons, so we have to pay attention to them after all. (Dennett 2005, 19–20)

The mechanistic framework I espouse in this book is also a kind of broad functionalism. It acknowledges the significance of multiple levels of organization and admits that different tools are needed to describe cognitive mechanisms. This naturally leads to a more pluralist approach to methodology; admittedly, it cannot make much sense of one-sided attempts to retract to one favorite kind of modeling strategy using just one representational format and one research strategy. Real science is messy. This, however, is not a bug; this is a feature.

Notes

1   Computation in Cognitive Science: Four Case Studies and a Funeral

1. I review just a couple, but there are more on the market of ideas. For example, in 2006, the European Network for the Advancement of Artificial Cognitive Systems (euCognition) asked new members of the network to submit an answer to the question of what cognition is. Out of around forty responses, one could not find two answers that were exactly equivalent (even if there were some similarities). I will return to the question of defining cognition in the concluding remarks of chapter 3.

2. The exact level of overregularization is somewhat contentious a matter (see Brooks et al. 1999). Bechtel and Abrahamsen (2002, section 5.4.1) summarize some of the methodological problems with estimating frequencies.

3. Wickelgren developed a context-free associative theory of speech encoding, but there was no decisive evidence of this theory being psychologically realistic.

2   Computational Processes

1. Forcefully defended by Jerry Fodor and sometimes conflated with the physical-symbol system hypothesis as championed by Newell and Simon (Newell 1980).

2. Note that Cantwell Smith seems to have abandoned this claim in his forthcoming book *Age of Significance* (as seen in his Introduction at http://www.ageofsignificance .org/aos/en/aos-v1c0.html, visited on March 6th, 2012). I thank Oron Shagrir for pointing my attention to this change of opinion.

3. If you think that ENIAC was still programmable, consider a possible copy of ENIAC whose wires were soldiered once and for all. Would the copy be a computer?

4. Another argument for the semantic view, usually voiced by the proponents of connectionism, is that at least some systems are computational in virtue of their

structural resemblance to task domains that they negotiate successfully (O'Brien and Opie 2006). In particular, during training, connectionist networks acquire resemblance to their task domains, and this is why "computations are causal processes that implicate one or more representing vehicles, such that their trajectory is shaped by the representational contents of those vehicles" (ibid., 32). But this argument is too weak; one would have to show that all computational systems are this way. Unfortunately for O'Brien and Opie, one can always do this using Searle's trick. In their account, structural resemblance is a (second-order) simple mapping relation between relations that obtain among connection weights and relations in the task domain. But reliance on simple mapping makes their view useless for dealing with the problem of spurious ascriptions of computation (after all, one can devise Cambridge-style logical relations in any domain).

Still another argument was used by Cummins and Schwartz (1991) who were aware of the possible trivialization by spurious ascriptions. They claim that physical systems compute if, and only if, they satisfy a computational function (which they take to be a matter of a simple mapping) and are semantically interpretable. Unfortunately, physical systems are always semantically interpretable if there are no special constraints on interpretation. Cummins and Schwartz cite no such constraints.

To sum up, these variants of the semantic view do not assume that there are atomic representations over which computation is defined. But they cannot deal with Searle's trivialization arguments.

5. One might retort that there is denotational semantics for Pascal, so this piece of code obviously denotes as a whole (for a similar argument, see Rescorla 2012). But formal semantics, denotational or not, for a programming language does not furnish it with genuine representational capacities at all. It simply specifies how the language is to be conventionally interpreted by cognitive agents. This interpretation however is not mandatory. Rescorla seems to admit this much by saying that only some of the computations are, in his terminology, *semantically permeable* (ibid., 11). But then semantics cannot be an essential feature of all computation (on the pain of contradiction).

6. MacKay uses "representation" instead of "elements" here, but this might be misleading, as no semantic property is presupposed (or denied) here. I resort to the neutral term "element" to avoid this kind of confusion; I reserve the word "representation" to refer to semantically loaded elements (see chapter 4).

7. Even if you claim that all physical systems are computational (a claim sometimes labeled pancomputationalism, cf. Piccinini 2007b), you still have to account for the difference between desktop computers, minds, and similar systems, and everything else, unless you want to make computational theory of mind trivial (see Miłkowski 2007). A pancomputationalist might think that high-level cognitive processing is implemented by computational processes rather than by merely physical processes,

but he or she must have a narrower concept of higher-level computational processing either way.

8. The mechanistic theory of implementation might also rely on other accounts of function provided that they fulfill certain criteria. It is not essential for my theory of implementation that the function is specified in terms of design, though this account fits quite easy with mechanistic explanation. Other accounts, such as those based on self-organizational principles (Bickhard 2008; Collier 2000) seem to be equally fine, though showing how they apply to artifact functionality without relying on designer's intention would require a bit more space. (Artifacts, at least the ones that are not self-organized or autonomous, need to derive their functionality from another functional process that is itself a part of a self-organized or autonomous entity.)

9. Yet the mechanistic theory of implementation does not rely on his theory in particular; it is simply one theory among many that fits—not as a token but as a type. In other words, it might be freely replaced by one's favorite account of function. As long the account warrants causal relevance of function and has a proper distinction between functioning and malfunctioning, it could be used instead.

10. It is easy to imagine an example of an unstable state of a pail of water that is almost falling (addition of more water would cause it to fall, etc.). However, the system behaviors above are the ones that are stably displayed by this kind of system.

11. Note however that the difference in computation is indiscernible only when we define it on the level of input and output relationships. As soon as we include, for example, exact state transitions of a machine, these computations will be different.

12. The search engine for the CiteSeerX database of the publications on information technology (developed at the University of Pennsylvania and hosted at http://citeseerx.ist.psu.edu) shows that the term was used in over 600 thousand papers; "implementation" appeared in publications about 500 thousand times, "algorithm" about 400 thousand times, and "information" about 750 thousand times. (Note that this is not a keyword frequency list, which is not easily available—just an informal inquiry.)

3   Computational Explanation

1. As Paweł Grabarczyk pointed out to me, another way to defend the CL account is to replace classical logic with a nonmonotonic one, as the counterexamples are mostly targeted at the material implication. For one proposal on how to use nonmonotonic logic in the CL theory, see Janssen and Tan (1991). However, the nonmonotonic version of the CL model would be formally much more complex. A simpler solution to some of the counterexamples would be simply to require that

the premises of the argument be the minimal set that implies the conclusion (the explanandum).

2. One could try to defend functionalist idealization by appeal to multiple models (MMI). Yet the proponents of functional analysis do not assert that multiple incompatible models are to be developed to explain the phenomenon, so it would go against their intentions, even if it could save some of the modeling efforts as reasonable.

3. Note that Newell and Dennett talk about cognitive systems exclusively, whereas Marr proposed his levels as an account of all computational systems (not just cognitive ones). I do not suggest that these accounts are equivalent; Marr's has the advantage of greater generality.

4. There is controversy over the explanatory value of competence theories as such. Some writers claim, for example, that such theories are too idealized (Franks 1995) to feature in three-level explanations of the kind envisaged by Marr. "Competence," it might be replied, is merely a name for a system that stores knowledge, which need not be idealized (Patterson 1998). This suggestion may not help Chomsky, however, as it implies that the competence module is redundant in the cognitive system (see Johnson-Laird 1987).

5. There are abuses of neurological evidence when it is supposed to replace the complete model of the capacity (Trout 2008; cf. Weisberg et al. 2008); such explanations are not considered correct according to the mechanistic account.

4 Computation and Representation

1. Fodor relied on the presupposition that explanation must proceed by appeal to the laws of nature, which, in turn, must be couched in terms of natural kinds (the covering-law account). His argument was that our representations can be about things, such as phlogiston or pencils, that are not natural kinds. Now, in the case of phlogiston, there is a causal explanation of why the representation fails to refer: phlogiston does not exist. As to the second case, the proposition that there are no causal explanations involving pencils is preposterous; for instance, a forensic expert might refer to a pencil in a detailed causal account of a suicide. What this means is that Fodor's concept of causation is too narrow. (The interventionist view does not preclude artifacts from being causal factors.)

2. Action, however, is not the most important factor here. For example, Taddeo and Floridi's (2007) Action-based Semantics (AbS) relies on the immediate current action, rather than the readiness to act, as the meaning of symbols to be grounded. In other words, what they mean by "actions" seems to be the same as good old behaviorists' reactions to stimuli. But as MacKay (1969) insists, it is not the reaction of the receiver that constitutes the meaning of information; otherwise, the receiver would have to

react in exactly the same way to the same information every time. Moreover, it makes nonutilized content devoid of meaning. This is highly undesirable. Imagine an animal that does not react to a perceived food stimulus because there is also a predator lurking near the food. The animal is ready to act, but we should not say that it did not represent the food just because it did not reach for it. Moreover, there is no representational system-detectable error in AbS, which is why the AbS model does not qualify as a correct account of representational mechanism.

3. Note, however, that the motor cortex is directly activated by low-level vision before it activates any of the two visual pathways (Goslin et al. 2012).

4. The very notion of first-person access or acquaintance strikes me as anything but clear. Why should the surface grammar of declarative sentences be important in defining something so crucial about cognition? Is it really only the first-person form of the verb that is required? Do plural sentences, such as "We know carrots," count as expressing first-person acquaintance with carrots? What about sentences such as "This color looks red to me," which, despite being phrased in the third person, seem to express some proposition about color perception? Clearly, then, the expression "first-person access/acquaintance" makes little sense when taken literally; I also suspect that any reasonable attempt to flesh out this metaphor would amount to replacing the grammatical term "first-person" with the traditional philosophical term "subjective." The meaning of this latter term is equally unclear, but at least it does not imply the existence of some deep truth about grammar.

5 Limits of Computational Explanation

1. The Polish translation was *Mały John szukał jego pudełko z zabawkami. W końcu go znalazł. Pudełko było w zagrodzie. John był bardzo szczęśliwy*. The machine translation system did not translate *pen* as *writing utensil* but as *enclosure* (*zagroda*); this would be impossible according to Bar-Hillel (1964), who wrote: "I now claim that no existing or imaginable program will enable an electronic computer to determine that the word *pen* in the given sentence within the given context has the second of the above meanings" (i.e., enclosure for small children). The translation is not perfect, as it should rather be *kojec*, but it is somewhat close to it. Anyway, such ambiguities might be statistically quite irrelevant, and human translators are prone to making the same mistakes as Google Translate (sometimes they fare even worse).

References

Adams, F. 2010. Why we still need a mark of the cognitive. *Cognitive Systems Research* 11 (4):324–331.

Andersen, H. K. 2011. Mechanisms, laws, and regularities. *Philosophy of Science* 78 (2):325–331.

Anderson, J. R. 2007. *How Can the Mind Occur in the Physical Universe?* Oxford: Oxford University Press.

Anderson, M. L., and G. Rosenberg. 2008. Content and action: The guidance theory of representation. *Journal of Mind and Behavior* 29 (1–2):55–86.

Aizawa, K. 2003. *The Systematicity Arguments*. Boston: Kluwer Academic.

Akins, K. 1996. Of sensory systems and the "aboutness" of mental states. *Journal of Philosophy* 93 (7):337–372.

Ariew, A., R. Cummins, and M. Perlman, eds. 2002. *Functions: New Essays in the Philosophy of Psychology and Biology*. Oxford: Oxford University Press.

Baars, B. J. 1986. *The Cognitive Revolution in Psychology*. New York: Guilford Press.

Baeza-Yates, R., and B. Ribeiro-Neto. 1999. *Modern Information Retrieval*. New York: ACM Press, Addison-Wesley.

Barandiaran, X., and A. Moreno. 2006. On what makes certain dynamical systems cognitive: A minimally cognitive organization program. *Adaptive Behavior* 14 (2):171–185.

Barandiaran, X., and A. Moreno. 2008. Adaptivity: From metabolism to behavior. *Adaptive Behavior* 16 (5):325–344.

Bar-Hillel, Y. 1964. A demonstration of the nonfeasibility of fully automatic high quality translation. In *Language and Information*, 174–179. Reading, MA: Addison-Wesley.

Barrett, H. C., and R. Kurzban. 2006. Modularity in cognition: Framing the debate. *Psychological Review* 113 (3):628–647.

Bartels, A. 2006. Defending the structural concept of representation. *Theoria* 21 (1):7–19.

Barwise, J., and J. Seligman. 1997. *Information Flow: The Logic of Distributed Systems*. Cambridge: Cambridge University Press.

Bateson, G. 1987. *Steps to an Ecology of Mind: Collected Essays in Anthropology, Psychiatry, Evolution, and Epistemology*. Northvale, NJ: Jason Aronson.

Bechtel, W. 1994. Levels of description and explanation in cognitive science. *Minds and Machines* 4 (1):1–25.

Bechtel, W. 1998. Representations and cognitive explanations: Assessing dynamicist's challenge in cognitive science. *Cognitive Science* 22 (3):295–318.

Bechtel, W. 2001. Representations and cognitive explanations: From neural systems to cognitive systems. In *Philosophy and the Neurosciences: A Reader*, 332–348, ed. W. Bechtel, P. Mandik, J. Mundale, and R. S. Stufflebeam. Oxford: Blackwell.

Bechtel, W. 2007. Reducing psychology while maintaining its autonomy via mechanistic explanation. In *The Matter of the Mind: Philosophical Essays on Psychology, Neuroscience, and Reduction*, ed. M. Schouten and H. Looren de Jong, 172–198. Oxford: Blackwell.

Bechtel, W. 2008a. *Mental Mechanisms*. New York: Routledge.

Bechtel, W. 2008b. Mechanisms in cognitive psychology: What are the operations? *Philosophy of Science* 75 (5):983–994.

Bechtel, W., and A. Abrahamsen. 2002. *Connectionism and the Mind*. Oxford: Blackwell.

Bechtel, W., and J. Mundale. 1999. Multiple realizability revisited: Linking cognitive and neural states. *Philosophy of Science* 66 (2):175–207.

Bechtel, W., and R. C. Richardson. 1993. *Discovering Complexity: Decomposition and Localization as Strategies in Scientific Research*. Princeton: Princeton University Press.

Beer, R. D. 1996. Toward the evolution of dynamical neural networks for minimally cognitive behavior. In *From animals to animats 11: Proceedings of the Fourth International Conference on Simulation of Adaptive Behavior*, ed. P. Maes, M. Mataric, J. A. Meyer, J. Pollack, and S. Wilson, (4):421–429. Cambridge, MA: MIT Press.

Beer, R. D. 2000. Dynamical approaches to cognitive science. *Trends in Cognitive Sciences* 4 (3):91–99.

Beer, R. D., and P. L. Williams. 2009. Animals and animats: Why not both iguanas? *Adaptive Behavior* 17 (4):296–302.

Bermúdez, J. L. 2010. *Cognitive Science: An Introduction to the Science of the Mind.* Cambridge: Cambridge University Press.

Bickhard, M. H. 1993. Representational content in humans and machines. *Journal of Experimental & Theoretical Artificial Intelligence* 5 (4):285–333.

Bickhard, M. H. 1998. Levels of representationality. *Journal of Experimental & Theoretical Artificial Intelligence* 10 (2):179–215.

Bickhard, M. H. 2007. Mechanism is not enough. *Pragmatics & Cognition* 15 (3):573–585.

Bickhard, M. H. 2008. The interactivist model. *Synthese* 166 (3):517–591.

Bickhard, M. H., and D. M. Richie. 1983. *On the Nature of Representation: A Case Study of James Gibson's Theory of Perception.* New York: Praeger.

Bickhard, M. H., and L. Terveen. 1995. *Foundational Issues in Artificial Intelligence and Cognitive Science: Impasse and Solution.* Amsterdam: North-Holland.

Bissell, C. 2007. Historical perspectives—The Moniac: A hydromechanical analog computer of the 1950s. *IEEE Control Systems Magazine* 27 (1):69–74.

Block, N. 1987. Advertisement for a semantics for psychology. *Midwest Studies in Philosophy* 10 (1):615–678.

Boden, M. A. 1988. *Computer Models of Mind: Computational Approaches in Theoretical Psychology.* Cambridge: Cambridge University Press.

Boden, M. A. 2004. *The Creative Mind: Myths and Mechanisms,* 2nd ed. London: Routledge.

Boden, M. A. 2008. Information, computation, and cognitive science. In *Philosophy of Information,* vol. 8, ed. P. Adriaans and J. Van Benthem, 741–761. Amsterdam: Elsevier B.V.

Bournez, O., and N. Dershowitz. Forthcoming. Foundations of analog algorithms. http://www.cs.tau.ac.il/~nachumd/papers/Analog.pdf.

Braitenberg, V. 1984. *Vehicles: Experiments in Synthetic Psychology.* Cambridge, MA: MIT Press.

Broadbent, D. E. 1958. *Perception and Communication.* Oxford: Pergamon Press.

Broadbent, D. E. 1985. A question of levels: Comment on McClelland and Rumelhart. *Journal of Experimental Psychology: General* 114:189–192.

Brooks, P. J., M. Tomasello, K. Dodson, and L. B. Lewis. 1999. Young children's overgeneralizations with fixed transitivity verbs. *Child Development* 70 (6):1325–1337.

Brown, T. H., E. W. Kairiss, and C. L. Keenan. 1990. Hebbian synapses: Biophysical mechanisms and algorithms. *Annual Review of Neuroscience* 13:475–511.

Burgess, N., J. G. Donnett, K. J. Jeffery, and J. O'Keefe. 1997. Robotic and neuronal simulation of the hippocampus and rat navigation. *Philosophical Transactions of the Royal Society of London, Series B: Biological Sciences* 352 (1360):1535–1543.

Burgess, N., J. G. Donnett, and J. O'Keefe. 1998. Using a mobile robot to test a model of the rat hippocampus. *Connection Science* 10 (3–4):291–300.

Burgess, N., A. Jackson, T. Hartley, and J. O'Keefe. 2000. Predictions derived from modelling the hippocampal role in navigation. *Biological Cybernetics* 83 (3):301–312.

Busemeyer, J. R., and A. Diederich. 2010. *Cognitive Modeling*. Los Angeles: Sage.

Bybee, J. L., and D. I. Slobin. 1982. Rules and schemas in the development and use of the English past tense. *Language* 58(2):265–289. doi:10.2307/414099.

Calvo, F. G. 2008. Towards a general theory of antirepresentationalism. *British Journal for the Philosophy of Science* 59 (3):259–292.

Calvo, P., and F. Keijzer. 2009. Cognition in plants. In *Plant-Environment Interactions: From Sensory Plant Biology to Active Plant Behavior*, ed. F. Baluška, 247–266. Berlin: Springer.

Cantwell Smith, B. 2002. The foundations of computing. In *Computationalism: New Directions*, ed. M. Scheutz. Cambridge, MA: MIT Press.

Cao, R. 2011. A teleosemantic approach to information in the brain. *Biology and Philosophy* 27 (1):49–71.

Cartwright, N. 1989. *Nature's Capacities and Their Measurement*. Oxford: Oxford University Press.

Cartwright, N. 2001. Modularity: It can—and generally does—fail. In *Stochastic Causality*, 65–84, ed. M. Galavotti, P. Suppes, and D. Costantini. Stanford: CSLI Publications.

Cartwright, N. 2002. Against modularity, the causal Markov condition, and any link between the two: Comments on Hausman and Woodward. *British Journal for the Philosophy of Science* 53:411–453.

Cartwright, N. 2004. Causation: One word, many things. *Philosophy of Science* 71:S805–S819.

Casacuberta, D., S. Ayala, and J. Vallverdu. 2010. Embodying cognition: A morphological perspective. In *Thinking Machines and the Philosophy of Computer Science*, ed. J. Vallverdú, 344–366. IGI Global.

Chaitin, G. J. 1990. *Information, Randomness, and Incompleteness: Papers on Algorithmic Information Theory*. Singapore: World Scientific.

Chalmers, D. J. 1993. Connectionism and compositionality: Why Fodor and Pylyshyn were wrong. *Philosophical Psychology* 6 (3):305–319.

Chalmers, D. J. 1996a. Does a rock implement every finite-state automaton? *Synthese* 108 (3):309–333.

Chalmers, D. J. 1996b. *The Conscious Mind: In Search of a Fundamental Theory*. Oxford: Oxford University Press.

Chalmers, D. J. 2011. A computational foundation for the study of cognition. *Journal of Cognitive Science* 12:325–359.

Chemero, A. 2000. Anti-representationalism and the dynamical stance. *Philosophy of Science* 67 (4):625–647.

Chomsky, N. 1959. Review of *Verbal Behavior* by B. F. Skinner. *Language* 35 (1):26–58.

Chrisley, R. 2000. Transparent computationalism. In *New Computationalism: Conceptus-Studien 14*, ed. M. Scheutz, 105–121. Sankt Augustin: Academia.

Churchland, P. M. 1996. *The Engine of Reason, The Seat of the Soul*. Cambridge, MA: MIT Press.

Churchland, P. S. 1989. *Neurophilosophy: Toward a Unified Science of the Mind-Brain*. Cambridge, MA: MIT Press.

Cisek, P. 1999. Beyond the computer metaphor: Behavior as interaction. *Journal of Consciousness Studies* 6 (11–12):125–142.

Clark, A. 1990. Connectionism, competence, and explanation. *British Journal for the Philosophy of Science* 41 (2):195.

Clark, A. 1997. *Being There: Putting Brain, Body, and World Together Again*. Cambridge, MA: MIT Press.

Clark, A. 2001a. *Mindware: An Introduction to the Philosophy of Cognitive Science*. Oxford: Oxford University Press.

Clark, A. 2001b. Visual experience and motor action: Are the bonds too tight? *Philosophical Review* 110 (4):495–519.

Clark, A. 2008. *Supersizing the Mind*. Oxford: Oxford University Press.

Clark, A., and D. Chalmers. 1998. The extended mind. *Analysis* 58 (1):7–19.

Clark, A., and J. Toribio. 1994. Doing without representing? *Synthese* 101 (3):401–431.

Cleeremans, A., and R. M. French. 1996. From chicken squawking to cognition: Levels of description and the computational approach in psychology. *Psychologica Belgica* 36:1–28.

Collier, J. 1999. Causation is the transfer of information. In *Causation, Natural Laws, and Explanation*, ed. Howard Sankey, 279–331. Dordrecht: Kluwer.

Collier, J. 2000. Autonomy and process closure as the basis for functionality. *Annals of the New York Academy of Sciences* 901:280–290.

Collier, J. 2010. Information, causation, and computation. In *Information and Computation*, ed. Gordana Dodig-Crnkovic and Mark Burgin. Singapore: World Scientific.

Conklin, J., and C. Eliasmith. 2005. A controlled attractor network model of path integration in the rat. *Journal of Computational Neuroscience* 18 (2):183–203.

Copeland, B. J. 1996. What is computation? *Synthese* 108 (3):335–359.

Copeland, B. J. 2004. Hypercomputation: Philosophical issues. *Theoretical Computer Science* 317 (1–3):251–267.

Craik, K. 1943. *The Nature of Explanation*. Cambridge: Cambridge University Press.

Craver, C. F. 2001. Role functions, mechanisms, and hierarchy. *Philosophy of Science* 68 (1):53–74.

Craver, C. F. 2007. *Explaining the Brain: Mechanisms and the Mosaic Unity of Neuroscience*. Oxford: Oxford University Press.

Craver, C. F., and W. Bechtel. 2007. Top-down causation without top-down causes. *Biology and Philosophy* 22 (4):547–563.

Cummins, R. 1975. Functional analysis. *Journal of Philosophy* 72 (20):741–765.

Cummins, R. 1983. *The Nature of Psychological Explanation*. Cambridge, MA: MIT Press.

Cummins, R. 1996. *Representations, Targets, and Attitudes*. Cambridge, MA: MIT Press.

Cummins, R. 2000. "How does it work?" vs. "What are the laws?": Two conceptions of psychological explanation. In *Explanation and Cognition*, ed. F. Keil and R. Wilson, 117–145. Cambridge, MA: MIT Press.

Cummins, R., and G. Schwarz. 1991. Connectionism, computation, and cognition. In *Connectionism and the Philosophy of Mind*, ed. T. Horgan and J. Tienson, 9:60–73. Dordrecht: Kluwer Academic.

Cutland, N. J. 1980. *Computability: An Introduction to Recursive Function Theory*. Cambridge: Cambridge University Press.

D'Mello, S., and S. Franklin. 2011. Computational modeling/cognitive robotics complements functional modeling/experimental psychology. *New Ideas in Psychology* 29 (3):217–227.

Daugman, J. G. 1990. Brain metaphor and brain theory. In *Computational Neuroscience*, ed. E. Schwartz, 9–18. Cambridge, MA: MIT Press.

Davidson, D. 1987. Knowing one's own mind. *Proceedings and Addresses of the American Philosophical Association* 60 (3):441–458.

Dawson, M. R. W. 2004. *Minds and Machines: Connectionism and Psychological Modeling*. Malden, MA: Blackwell.

Dayan, P., and L. F. Abbott. 2001. *Theoretical Neuroscience: Computational and Mathematical Modeling of Neural Systems*. Cambridge, MA: MIT Press.

Dennett, D. C. 1969. *Content and Consciousness*. London: Routledge & Kegan Paul.

Dennett, D. C. 1978. *Brainstorms: Philosophical Essays on Mind and Psychology*. Cambridge, MA: MIT Press.

Dennett, D. C. 1987. *The Intentional Stance*. Cambridge, MA: MIT Press.

Dennett, D. C. 1991a. Ways of establishing harmony. In *Dretske and His Critics*, ed. B. P. McLaughlin, 118–130. Malden, MA: Blackwell.

Dennett, D. C. 1991b. Real patterns. *Journal of Philosophy* 88 (1):27–51.

Dennett, D. C. 1995. *Darwin's Dangerous Idea: Evolution and the Meaning of Life*. New York: Simon & Schuster.

Dennett, D. C. 1996. In defense of AI. In *Speaking Minds: Interviews with Twenty Eminent Cognitive Scientists*, ed. P. Baumgartner and S. Payr, 59–69. Princeton: Princeton University Press.

Dennett, D. C. 1998. *Brainchildren: Essays on Designing Minds*. Cambridge, MA: MIT Press.

Dennett, D. C. 2005. *Sweet Dreams: Philosophical Obstacles to a Science of Consciousness*. Cambridge, MA: MIT Press.

Derdikman, D., and E. I. Moser. 2010. A manifold of spatial maps in the brain. *Trends in Cognitive Sciences* 14 (12):561–569.

Deutsch, D. 1985. Quantum theory, the Church–Turing principle, and the universal quantum computer. *Proceedings of the Royal Society of London, Series A: Mathematical and Physical Sciences* 400 (1818):97–117.

Devitt, M. 1991. Why Fodor can't have it both ways. In *Meaning in Mind: Fodor and His Critics*, ed. B. Loewer and G. Rey, 95–118. Oxford: Blackwell.

Dienes, Z., and J. Perner. 2007. Executive control without conscious awareness: The cold control theory of hypnosis. In *Hypnosis and Conscious States: The Cognitive Neuroscience Perspective*, ed. G. Jamieson, 293–314. Oxford: Oxford University Press.

Dietrich, E. 2007. Representation. In *Philosophy of Psychology and Cognitive Science*, ed. P. Thagard, 1–29. Amsterdam: North-Holland.

Douglas, H. E. 2009. Reintroducing prediction to explanation. *Philosophy of Science* 76 (October):444–463.

Drescher, G. L. 1991. *Made-up Minds: A Constructivist Approach to Artificial Intelligence.* Cambridge, MA: MIT Press.

Dresner, E. 2010. Measurement-theoretic representation and computation-theoretic realization. *Journal of Philosophy* 107 (6):275–292.

Dretske, F. 1982. *Knowledge and the Flow of Information,* 2nd ed. Cambridge, MA: MIT Press.

Dretske, F. 1986. Misrepresentation. In *Belief: Form, Content, and Function,* ed. R. J. Bogdan, 17–36. Oxford: Clarendon Press.

Dreyfus, H. 1972. *What Computers Can't Do: A Critique of Artificial Reason.* New York: Harper & Row.

van Duijn, M., F. Keijzer, and D. Franken. 2006. Principles of minimal cognition: Casting cognition as sensorimotor coordination. *Adaptive Behavior* 14 (2):157–170.

Eliasmith, C. 2003. Moving beyond metaphors: Understanding the mind for what it is. *Journal of Philosophy* 100 (10):493–520.

Eliasmith, C. 2005a. A unified approach to building and controlling spiking attractor networks. *Neural Computation* 17 (6):1276–1314.

Eliasmith, C. 2005b. Cognition with neurons: A large-scale, biologically realistic model of the Wason task. In *Proceedings of the XXVII Annual Conference of the Cognitive Science Society,* ed. G. Bara, L. Barsalou, and M. Bucciarelli. Stressa, Italy.

Eliasmith, C. 2009. Neurocomputational models: Theory, application, philosophical consequences. In *The Oxford Handbook of Philosophy and Neuroscience,* ed. J. Bickle, 346–369. New York: Oxford University Press.

Eliasmith, C. 2010. How we ought to describe computation in the brain. *Studies in History and Philosophy of Science Part A* 41 (3):313–320.

Eliasmith, C., and C. H. Anderson. 2003. *Neural Engineering: Computation, Representation, and Dynamics in Neurobiological Systems.* Cambridge, MA: MIT Press.

Elman, J. L. 1990. Finding structure in time. *Cognitive Science* 14 (2):179–211.

Elman, J. L. 1991. Distributed representations, simple recurrent networks, and grammatical structure. *Machine Learning* 7 (2–3):195–225.

Farrell, S., and S. Lewandowsky. 2010. Computational models as aids to better reasoning in psychology. *Current Directions in Psychological Science* 19 (5):329–335.

Fischer, B. J., J. L. Peña, and M. Konishi. 2007. Emergence of multiplicative auditory responses in the midbrain of the barn owl. *Journal of Neurophysiology* 98 (3): 1181–1193.

Floridi, L. 2008a. Trends in the philosophy of information. In *Philosophy of Information*, vol. 8, ed. P. Adriaans and J. Van Benthem, 113–131. Amsterdam: Elsevier B.V.

Floridi, L. 2008b. The method of levels of abstraction. *Minds and Machines* 18 (3):303–329.

Fodor, J. A. 1968. *Psychological Explanation: An Introduction to the Philosophy of Psychology*. New York: Random House.

Fodor, J. A. 1974. Special sciences (or: The disunity of science as a working hypothesis). *Synthese* 28 (2):97–115.

Fodor, J. A. 1975. *The Language of Thought*. New York: Thomas Y. Crowell.

Fodor, J. A. 1980. Methodological solipsism considered as a research strategy in cognitive psychology. *Behavioral and Brain Sciences* III (1):63–72.

Fodor, J. A. 1983. *The Modularity of Mind*. Cambridge, MA: MIT Press.

Fodor, J. A. 2000. *The Mind Doesn't Work That Way: The Scope and Limits of Computational Psychology*. Cambridge, MA: MIT Press.

Fodor, J., and E. Lepore. 1993. *Holism: A Shopper's Guide*. Oxford: Blackwell.

Fodor, J., and Z. W. Pylyshyn. 1988. Connectionism and cognitive architecture: A critical analysis. *Cognition* 28:3–71.

Forbus, K. D. 2010. AI and cognitive science: The past and next 30 years. *Topics in Cognitive Science* 2 (3):345–356.

Franklin, S., and F. G. Patterson Jr. 2006. The LIDA architecture: Adding new modes of learning to an intelligent, autonomous, software agent. *Integrated Design and Process Technology* 703:764–1004.

Franks, B. 1995. On explanation in the cognitive sciences: Competence, idealization, and the failure of the classical cascade. *British Journal for the Philosophy of Science* 46 (4):475–502.

Fresco, N. 2008. An analysis of the criteria for evaluating adequate theories of computation. *Minds and Machines* 18 (3):379–401.

Fresco, N. 2010. Explaining computation without semantics: Keeping it simple. *Minds and Machines* 20 (2):165–181.

Frijda, N. H. 1967. Problems of computer simulation. *Behavioral Science* 12 (1):59–67.

Gallistel, C. R. 1990. *The Organization of Learning*. Cambridge, MA: MIT Press.

Giunti, M. 1997. *Computation, Dynamics, and Cognition*. New York: Oxford University Press.

Glennan, S. S. 1996. Mechanisms and the nature of causation. *Erkenntnis* 44 (1):49–71.

Glennan, S. S. 2002. Rethinking mechanistic explanation. *Philosophy of Science* 69 (no. S3):S342–S353.

Glennan, S. S. 2005. Modeling mechanisms. *Studies in History and Philosophy of Science, Part C: Studies in History and Philosophy of Biological and Biomedical Sciences* 36 (2):443–464.

Godfrey-Smith, P. 1989. Misinformation. *Canadian Journal of Philosophy* 19 (4):533–550.

Godfrey-Smith, P. 2008. Triviality arguments against functionalism. *Philosophical Studies* 145 (2):273–295.

Goslin, J., T. Dixon, M. H. Fischer, A. Cangelosi, and R. Ellis. 2012. Electrophysiological examination of embodiment in vision and action. *Psychological Science* 23 (2):152–157.

Griffiths, T. L., N. Chater, C. Kemp, A. Perfors, and J. B. Tenebaum. 2010. Probabilistic models of cognition: Exploring representations and inductive biases. *Trends in Cognitive Sciences* 14 (8):357–364.

Gross, C. G. 2002. Genealogy of the "grandmother cell." *Neuroscientist* 8 (5): 512–518.

Grünwald, P. D. 2007. *The Minimum Description Length Principle*. Cambridge, MA: MIT Press.

Grush, R. 2003. In defense of some "Cartesian" assumptions concerning the brain and its operation. *Biology and Philosophy* 18:53–93.

Grush, R. 2004. The emulation theory of representation: Motor control, imagery, and perception. *Behavioral and Brain Sciences* 27 (3):377–396.

Guazzelli, A., M. Bota, F. J. Corbacho, and M. A. Arbib. 1998. Affordances, motivations, and the world graph theory. *Adaptive Behavior* 6 (3–4):435–471.

Gurevich, Y. 1995. Evolving algebras 1993: Lipari guide. In *Specification and Validation Methods*, ed. E. Börger, 231–243. Oxford: Oxford University Press.

Hardcastle, V. G. 1996. *How to Build a Theory in Cognitive Science*. Albany: SUNY Press.

Harms, W. F. 1998. The use of information theory in epistemology. *Philosophy of Science* 65 (3):472–501.

Harnad, S. 1990. The symbol grounding problem. *Physica D: Nonlinear Phenomena* 42:335–346.

Hartmann, S. 1998. Idealization in quantum field theory. In *Idealization in Contemporary Physics*, ed. N. Shanks, 99–122. Amsterdam: Rodopi.

Haugeland, J. 1985. *Artificial Intelligence: The Very Idea*. Cambridge, MA: MIT Press.

Hempel, C. G., and P. Oppenheim. 1948. Studies in the logic of explanation. *Philosophy of Science* 15 (2):135–175.

Hofstadter, D.R., and D. C. Dennett. 1981. *The Mind's I: Fantasies and Reflections on Self and Soul*. New York: Bantam Dell.

Hunt, E. 1968. Computer simulation: Artificial intelligence studies and their relevance to psychology. *Annual Review of Psychology* 19 (1):135–168.

Hunt, E. 1989. Cognitive science: Definition, status, and questions. *Annual Review of Psychology* 40 (1):603–629.

Izhikevich, E. M. 2007. *Dynamical Systems in Neuroscience: The Geometry of Excitability and Bursting*. Cambridge, MA: MIT Press.

Jackson, F., and P. Pettit. 1990. Program explanation: A general perspective. *Analysis* 50 (2):107–117.

Jamieson, A. 1910. *Theory of Machines and Practical Mechanisms*, 7th ed., vol. 5: *A Text-book of Applied Mechanics and Mechanical Engineering*. London: Griffin.

Janssen, M. C. W., and Y.-H. Tan. 1991. Why Friedman's non-monotonic reasoning defies Hempel's covering law model. *Synthese* 86 (2):255–284.

Johnson-Laird, P. N. 1987. Grammar and psychology. In *Noam Chomsky: Consensus and Controversy*, ed. S. Modgil and C. Modgil, 147–156. London: Falmer Press.

Jones, M., and B. C. Love. 2011. Bayesian fundamentalism or enlightenment? On the explanatory status and theoretical contributions of Bayesian models of cognition. *Behavioral and Brain Sciences* 34 (04):169–188.

de Jong, H. L. 2003. Causal and functional explanations. *Theory & Psychology* 13 (3):291–317.

Kaplan, D. M., and C. F. Craver. 2011. The explanatory force of dynamical and mathematical models in neuroscience: A mechanistic perspective. *Philosophy of Science* 78 (4):601–627.

Kauffman, S. A. 1970. Articulation of parts explanation in biology and the rational search for them. In *PSA: Proceedings of the Biennial Meeting of the Philosophy of Science Association*, 257–272. Chicago: Philosophy of Science Association; University of Chicago Press.

Kim, J. 1993. *Supervenience and Mind: Selected Philosophical Essays*. Cambridge: Cambridge University Press.

Kirchhoff, M. D. 2011. Anti-representationalism: Not a well-founded theory of cognition. *Res Cogitans* 2:1–34.

Kitcher, P. 1988. Marr's computational theory of vision. *Philosophy of Science* 55 (1):1–24.

Kitcher, P. 1989. Explanatory unification and the causal structure of the world. In *Scientific Explanation*, vol. 13, ed. P. Kitcher and W. Salmon, 410–505. Minnesota Studies in the Philosophy of Science. Minneapolis: University of Minnesota Press.

Konorski, J. 1967. *Integrative Activity of the Brain: An Interdisciplinary Approach.* Chicago: University of Chicago Press.

Krajewski, S. 2007. On Gödel's theorem and mechanism: Inconsistency or unsoundness is unavoidable in any attempt to "out-Gödel" the mechanist. *Fundamenta Informaticae* 81 (1):173–181.

Kripke, S. A. 1982. *Wittgenstein on Rules and Private Language: An Elementary Exposition.* Cambridge, MA: Harvard University Press.

Krohs, U. 2007. Functions as based on a concept of general design. *Synthese* 166 (1):69–89.

Kuo, P. D., and C. Eliasmith. 2005. Integrating behavioral and neural data in a model of zebrafish network interaction. *Biological Cybernetics* 93 (3):178–187.

Kuorikoski, J. 2008. Varieties of modularity for causal and constitutive explanations. http://philsci-archive.pitt.edu/4303/.

Ladyman, J., D. Ross, D. Spurrett, and J. Collier. 2007. *Every Thing Must Go: Metaphysics Naturalized.* Oxford: Oxford University Press.

Larkin, J., and H. A. Simon. 1987. Why a diagram is (sometimes) worth ten thousand words. *Cognitive Science* 11 (1):65–100.

Lettvin, J. Y., H. R. Maturana, W. S. McCulloch, and W. H. Pitts. 1959. What the frog's eye tells the frog's brain. *Proceedings of the IRE* 47 (11):1940–1951.

Leuridan, B. 2010. Can mechanisms really replace laws of nature? *Philosophy of Science* 77 (3):317–340.

Levins, R. 1966. The strategy of model building in population biology. *American Scientist* 54 (4):421–431.

Lewandowsky, S. 1993. The rewards and hazards of computer simulations. *Psychological Science* 4 (4):236–243.

Lewandowsky, S., and S. Farrell. 2011. *Computational Modeling in Cognition: Principles and Practice.* Thousand Oaks: Sage Publications.

Litt, A., C. Eliasmith, and P. Thagard. 2008. Neural affective decision theory: Choices, brains, and emotions. *Cognitive Systems Research* 9 (4):252–273.

Liu, B., M. Caetano, N. Narayanan, C. Eliasmith, and M. Laubach. 2011. A neuronal mechanism for linking actions to outcomes in the medial prefrontal cortex. Poster presented at Computational and Systems Neuroscience 2011, Salt Lake City, Utah, February 24–27.

Lloyd, S. 2000. Ultimate physical limits to computation. *Nature* 406 (6799): 1047–1054.

Lloyd, S. 2002. Computational capacity of the universe. *Physical Review Letters* 88 (2):1–17.

Lucas, J. R. 1961. Minds, machines, and Gödel. *Philosophy* 9 (3):219–227.

Lund, H. H., B. Webb, and J. Hallam. 1997. A robot attracted to the cricket species *Gryllus Bimaculatus*. In *Fourth European Conference on Artificial Life*, ed. P. Husbands and I. Harvey. Cambridge, MA: MIT Press.

Lycan, W. G. 1987. *Consciousness*. Cambridge, MA: MIT Press.

MacDonald, J., and H. McGurk. 1978. Visual influences on speech perception processes. *Perception & Psychophysics* 24:253–257.

Machamer, P., L. Darden, and C. F. Craver. 2000. Thinking about mechanisms. *Philosophy of Science* 67 (1):1–25.

Machery, E. 2011. Why I stopped worrying about the definition of life . . . and why you should as well. *Synthese* 185 (1):145–164.

Mackay, D. M. 1951. Mindlike behaviour in artefacts. *British Journal for the Philosophy of Science* II (6):105–121.

MacKay, D. M. 1969. *Information, Mechanism, and Meaning*. Cambridge, MA: MIT Press.

MacNeil, D., and C. Eliasmith. 2011. Fine-tuning and the stability of recurrent neural networks. *PLoS ONE* 6 (9):e22885.

Mandik, P. 2005. Action-oriented representation. In *Cognition and the Brain: The Philosophy and Neuroscience Movement*, 284–305, ed. A. Brook and K. Akins. Cambridge: Cambridge University Press.

Mareschal, D., and M. S. C. Thomas. 2007. Computational modeling in developmental psychology. *IEEE Transactions on Evolutionary Computation* 11 (2):137–150.

Marr, D. 1982. *Vision: A Computational Investigation into the Human Representation and Processing of Visual Information*. New York: W. H. Freeman.

Marr, D., and T. Poggio. 1976. Cooperative computation of stereo disparity. *Science* 194 (4262):283–287.

Massaro, D. W., and N. Cowan. 1993. Information processing models: Microscopes of the mind. *Annual Review of Psychology* 44:383–425.

Maturana, H. R., and F. J. Varela. 1980. *Autopoiesis and Cognition: The Realization of the Living.* Dordrecht: Reidel.

McClamrock, R. A. 1991. Marr's three levels: A re-evaluation. *Minds and Machines* 1 (2):185–196.

McClamrock, R. A. 1995. *Existential Cognition: Computational Minds in the World.* Chicago: University of Chicago Press.

McClelland, J. L. 2002. "Words or rules" cannot exploit the regularity in exceptions: Reply to Pinker and Ullman. *Trends in Cognitive Sciences* 6 (11):464–465.

McClelland, J. L., and K. Patterson. 2002. Rules or connections in past-tense inflections: What does the evidence rule out? *Trends in Cognitive Sciences* 6 (11): 465–472.

McClelland, J. L., and D. E. Rumelhart, and the PDP Research Group, eds. 1986. *Parallel Distributed Processing: Explorations in the Microstructures of Cognition*, vol. 2: *Psychological and Biological Models.* Cambridge, MA: MIT Press.

McDermott, D. 2001. *Mind and Mechanism.* Cambridge, MA: MIT Press.

McNaughton, B. L., F. P. Battaglia, O. Jensen, E. I. Moser, and M.-B. Moser. 2006. Path integration and the neural basis of the "cognitive map." *Nature Reviews: Neuroscience* 7 (8):663–678.

Meyer, D. E., A. M. Osman, D. E. Irwin, and S. Yantis. 1988. Modern mental chronometry. *Biological Psychology* 26 (1–3):3–67.

Miłkowski, M. 2007. Is computationalism trivial? In *Computation, Information, Cognition—The Nexus and the Liminal*, ed. G. Dodig-Crnkovic and S. Stuart, 236–246. Newcastle: Cambridge Scholars Press.

Miłkowski, M. 2008. When weak modularity is robust enough? *Análisis Filosófico* 28 (1):77–89.

Miłkowski, M. 2010. Developing an open-source, rule-based proofreading tool. *Software, Practice & Experience* 40 (7):543–566.

Miłkowski, M. 2012. Is computation based on interpretation? *Semiotica* 15:229–238.

Miller, G. A. 1956. The magical number seven, plus or minus two: Some limits on our capacity for processing information. *Psychological Review* 63 (2):81–97.

Miller, G. A., E. Galanter, and K. H. Pribram. 1967. *Plans and the Structure of Behavior.* New York: Holt.

Millikan, R. G. 1984. *Language, Thought, and Other Biological Categories: New Foundations for Realism.* Cambridge, MA: MIT Press.

Millikan, R. G. 1989. Biosemantics. *Journal of Philosophy* 86 (6):281–297.

Millikan, R. G. 2002. Biofunctions: Two paradigms. In *Functions: New Essays in the Philosophy of Psychology and Biology*, ed. A. Ariew, R. Cummins, and M. Perlman. New York: Oxford University Press.

Milner, D., and M. Goodale. 1995. *The Visual Brain in Action*. Oxford: Oxford University Press.

Montebelli, A., R. Lowe, I. Ieropoulos, C. Melhuish, J. Greenman, and T. Ziemke. 2010. Microbial fuel cell driven behavioral dynamics in robot simulations. In *Artificial Life XII: Proceedings of the Twelfth International Conference on the Synthesis and Simulation of Living Systemsational Conference on the Synthesis and Simulation of Living Systems*, 749–756, ed. H. Fellermann, M. Dörr, M. M. Hanczyc, L. Ladegaard Laursen, S. Maurer, D. Merkle, P.-A. Monnard, K. Stoy, and S. Rasmussen. Cambridge, MA: MIT Press.

Moore, E. F. 1956. Gedanken-experiments on sequential machines. In *Automata Studies*, ed. C. E. Shannon and J. McCarthy, 129–153. Princeton: Princeton University Press.

Morse, A. F., C. Herrera, R. Clowes, A. Montebelli, and T. Ziemke. 2011. The role of robotic modelling in cognitive science. *New Ideas in Psychology* 29 (3):312–324.

Mundale, J., and W. Bechtel. 1996. Integrating neuroscience, psychology, and evolutionary biology through a teleological conception of function. *Minds and Machines* (6):481–505.

Nagy, N., and S. Akl. 2011. Computations with uncertain time constraints: Effects on parallelism and universality. In *Unconventional Computation*, 6714:152–163, ed. C. Calude, J. Kari, I. Petre, and G. Rozenberg. Berlin: Springer.

Neisser, U. 1963. The imitation of man by machine: The view that machines will think as man does reveals misunderstanding of the nature of human thought. *Science* 139 (3551):193–197.

Neisser, U. 1967. *Cognitive Psychology*. New York: Appleton-Century-Crofts.

Newell, A. 1973. You can't play 20 questions with nature and win: Projective comments on the papers of this symposium. In *Visual Information Processing*, 283–308, ed. W. G. Chase. New York: Academic Press.

Newell, A. 1980. Physical symbol systems. *Cognitive Science: A Multidisciplinary Journal* 4 (2): 135–183.

Newell, A. 1981. The knowledge level: Presidential address. *AI Magazine* 2 (2):1–21.

Newell, A. 1990. *Unified Theories of Cognition*. Cambridge, MA: Harvard University Press.

Newell, A., and H. A. Simon. 1972. *Human Problem Solving*. Englewood Cliffs, NJ: Prentice-Hall.

Newell, A., and H. A. Simon. 1976. Computer science as empirical inquiry: Symbols and search. *Communications of the ACM* 19 (3):113–126.

Nielsen, K. S. 2010. Representation and dynamics. *Philosophical Psychology* 23 (6):759–773.

Nowak, L. 1980. *The Structure of Idealization*. Dordrecht: Reidel.

Nowak, L. and I. Nowakowa. 2000. *Idealization X: The Richness of Idealization (Poznan Studies in the Philosophy of Sciences and the Humanities, 69)*. Amsterdam: Rodopi.

O'Brien, G., and J. Opie. 2006. How do connectionist networks compute? *Cognitive Processing* 7 (1):30–41.

O'Hara, K. 1994. Mind as machine: Can computational processes be regarded as explanatory of mental processes? Ph.D. Dissertation, University of Oxford.

O'Keefe, J., and L. Nadel. 1978. *The Hippocampus as a Cognitive Map*. Oxford: Clarendon Press.

O'Reilly, R. C. 2006. Biologically based computational models of high-level cognition. *Science* 314 (5796):91–94.

Palmer, S. E. 1978. Fundamental aspects of cognitive representation. In *Cognition and Categorization*, ed. E. Rosch and B. Lloyd, 259–303. Mahwah, NJ: Lawrence Erlbaum.

Paprzycka, K. 2005. *O możliwości antyredukcjonizmu*. Warszawa: Semper.

Parnas, D. L. 1972. On the criteria to be used in decomposing systems into modules. *Communications of the ACM* 15 (12):1053–1058.

Patterson, S. 1998. Discussion. Competence and the classical cascade: A reply to Franks. *British Journal for the Philosophy of Science* 49 (4):625–636.

Paul, C. 2006. Morphological computation: A basis for the analysis of morphology and control requirements. *Robotics and Autonomous Systems* 54 (8):619–630.

Peacocke, C. 1986. Explanation in computational psychology: Language, perception, and level 1.5. *Mind & Language* 1 (2):101–123.

Pearl, J. 2000. *Causality: Models, Reasoning, and Inference*. Cambridge: Cambridge University Press.

Penrose, R. 1989. *The Emperor's New Mind*. London: Oxford University Press.

Petty, M. D. 2010. Verification, validation, and accreditation. In *Modeling and Simulation Fundamentals*, 325–372, ed. J. A. Sokolowski and C. M. Banks. New York: John Wiley & Sons.

Pfeifer, R., and F. Iida. 2005. Morphological computation: Connecting body, brain, and environment. *Japanese Scientific Monthly* 58 (2):48–54.

Pfeifer, R., F. Iida, and G. Gomez. 2006. Morphological computation for adaptive behavior and cognition. *International Congress Series* 1291:22–29.

Piasecki, M., S. Szpakowicz, and B. Broda. 2009. *A Wordnet from the Ground Up*. Wrocław: Oficyna Wydawnicza Politechniki Wrocławskiej.

Piccinini, G. 2006. Computation without representation. *Philosophical Studies* 137 (2):205–241.

Piccinini, G. 2007a. Computing mechanisms. *Philosophy of Science* 74 (4): 501–526.

Piccinini, G. 2007b. Computational modelling vs. computational explanation: Is everything a Turing machine, and does it matter to the philosophy of mind? *Australasian Journal of Philosophy* 85 (1):93–115.

Piccinini, G. 2008. Computers. *Pacific Philosophical Quarterly* 89:32–73.

Piccinini, G. 2010. Computation in physical systems. In *The Stanford Encyclopedia of Philosophy*, fall 2010 edition, ed. Edward N. Zalta. http://plato.stanford.edu/entries/computation-physicalsystems/.

Piccinini, G., and S. Bahar. Forthcoming. Neural computation and the computational theory of cognition. *Cognitive Science*.

Piccinini, G., and C. Craver. 2011. Integrating psychology and neuroscience: Functional analyses as mechanism sketches. *Synthese* 183 (3):283–311.

Piccinini, G., and A. Scarantino. 2011. Information processing, computation, and cognition. *Journal of Biological Physics* 37 (1):1–38.

Pinker, S., and A. Prince. 1988. On language and connectionism: Analysis of a parallel distributed processing model of language acquisition. *Cognition* 23:73–193.

Pinker, S., and M. T. Ullman. 2002a. The past-tense debate: The past and future of the past tense. *Trends in Cognitive Sciences* 6 (11):456–463.

Pinker, S., and M. Ullman. 2002b. Combination and structure, not gradedness, is the issue. *Trends in Cognitive Sciences* 6 (11):472–474.

Pitt, M. A., I. J. Myung, and S. Zhang. 2002. Toward a method of selecting among computational models of cognition. *Psychological Review* 109 (3):472–491.

Polanyi, M. 1958. *Personal Knowledge: Towards a Post-critical Philosophy*. Chicago: University of Chicago Press.

Polger, T. W. 2004. Neural machinery and realization. *Philosophy of Science* 71 (5):997–1006.

Polger, T. W. 2008. Evaluating the evidence for multiple realization. *Synthese* 167 (3):457–472.

Posner, M. I. 2005. Timing the brain: Mental chronometry as a tool in neuroscience. *PLoS Biology* 3 (2):e51. doi:10.1371/journal.pbio.0030051.

Posner, M. I., and P. McLeod. 1982. Information processing models–in search of elementary operations. *Annual Review of Psychology* 33:477–514.

Prinz, J. 2000. The ins and outs of consciousness. *Brain and Mind* 1:245–256.

Putnam, H. 1960. Minds and machines. In *Dimensions of Mind*, ed. S. Hook. New York University Press.

Putnam, H. 1991. *Representation and Reality*. Cambridge, MA: MIT Press.

Pylyshyn, Z. W. 1984. *Computation and Cognition: Toward a Foundation for Cognitive Science*. Cambridge, MA: MIT Press.

Pylyshyn, Z. W. 1989. Computing in cognitive science. In *Foundations of Cognitive Science*, ed. M. Posner, 51–91. Cambridge, MA: MIT Press.

Raatikainen, P. 2010. Causation, exclusion, and the special sciences. *Erkenntnis* 73 (3):349–363.

Ramsey, W. M. 2007. *Representation Reconsidered*. Cambridge: Cambridge University Press.

Rapaport, W. J. 1998. How minds can be computational systems. *Journal of Experimental & Theoretical Artificial Intelligence* 10 (4):403–419.

Redish, A. D. 1999. *Beyond the Cognitive Map: From Place Cells to Episodic Memory*. Cambridge, MA: MIT Press.

Reiter, R. 2001. *Knowledge in Action: Logical Foundations for Specifying and Implementing Dynamical Systems*. Cambridge, MA: MIT Press.

Rescorla, M. 2012. How to integrate representation into computational modeling, and why we should. *Journal of Cognitive Science* 13 (1):1–38.

Rolls, E. T. 2007. *Memory, Attention, and Decision-Making: A Unifying Computational Neuroscience Approach*. Oxford: Oxford University Press.

Rolls, E. T., and A. Treves. 2011. The neuronal encoding of information in the brain. *Progress in Neurobiology* 95 (3):448–490.

Rosenblatt, F. 1958. The perceptron: A probabilistic model for information storage and organization in the brain. *Psychological Review* 65 (6):386–408.

Ross, D., and D. Spurrett. 2004. What to say to a skeptical metaphysician: A defense manual for cognitive and behavioral scientists. *Behavioral and Brain Sciences* 27 (5): 603–627; discussion 627–647.

Rowlands, M. 2009. Extended cognition and the mark of the cognitive. *Philosophical Psychology* 22 (1):1–19.

Ruben, D. H. 1992. *Explaining Explanation*. London: Routledge.

Rumelhart, D. E., and J. L. McClelland. 1985. Levels indeed! A response to Broadbent. *Journal of Experimental Psychology: General* 114:193–197.

Rumelhart, D. E., and J. L. McClelland. 1986. On learning the past tenses of English verbs. In *Parallel Distributed Processing: Explorations in the Microstructures of Cognition*, vol. 2: *Psychological and Biological Models*, ed. J. L. McClelland, D. E. Rumelhart, and the PDP Research Group, 216–271. Cambridge, MA: MIT Press.

Rupert, R. D. 2009. *Cognitive Systems and the Extended Mind*. Oxford: Oxford University Press.

Rusanen, A. M., and O. Lappi. 2007. The limits of mechanistic explanation in neurocognitive sciences. In *Proceedings of the European Cognitive Science Conference 2007*. Mahwah, NJ: Lawrence Erlbaum.

Samsonovich, A., and B. L. McNaughton. 1997. Path integration and cognitive mapping in a continuous attractor neural network model. *Journal of Neuroscience: The Official Journal of the Society for Neuroscience* 17 (15): 5900–5920.

Scheutz, M. 1996. When physical systems realize functions. . . . *Minds and Machines* 9 (2):1–34.

Scheutz, M. 2000. The ontological status of representations. In *Understanding Representation in the Cognitive Sciences*, 33–38, ed. A. Riegler, M. Peschl, and A. Stein. Berlin: Springer.

Scheutz, M. 2001. Computational versus causal complexity. *Minds and Machines* 11:543–566.

Schulte, P. 2012. How frogs see the world: Putting Millikan's teleosemantics to the test. *Philosophia*. doi:10.1007/s11406-011-9358-x.

Searle, J. R. 1980. Minds, brains, and programs. *Behavioral and Brain Sciences* 3:1–19.

Searle, J. R. 1984. *Minds, Brains, and Science*. Cambridge, MA: Harvard University Press.

Searle, J. R. 1992. *The Rediscovery of Mind*. Cambridge, MA: MIT Press.

Sekiyama, K., and Y. Tohkura. 1991. McGurk effect in non-English listeners: Few visual effects for Japanese subjects hearing Japanese syllables of high auditory intelligibility. *Journal of the Acoustical Society of America* 90 (4):1797–1805.

Shagrir, O. 2006. Why we view the brain as a computer. *Synthese* 153 (3): 393–416.

Shagrir, O. 2010a. Marr on computational-level theories. *Philosophy of Science* 77 (4):477–500.

Shagrir, O. 2010b. Computation, San Diego style. *Philosophy of Science* 77 (5): 862–874.

Shagrir, O. 2010c. Towards a modeling view of computing. In *Information and Computation*, ed. G. Dodig-Crnkovic and M. Burgin. Singapore: World Scientific Publishing.

Shagrir, O. 2010d. Brains as analog-model computers. *Studies In History and Philosophy of Science, Part A* 41 (3):271–279.

Shanahan, M. 1997. *Solving the Frame Problem: A Mathematical Investigation of the Common Sense Law of Inertia*. Cambridge, MA: MIT Press.

Shanahan, M., and B. Baars. 2005. Applying global workspace theory to the frame problem. *Cognition* 98 (2):157–176.

Shannon, C. E. 1948. A mathematical theory of communication. *Bell System Technical Journal* 27:379–423, 623–656.

Shapiro, L. A. 2004. *The Mind Incarnate*. Cambridge, MA: MIT Press.

Shapiro, L. A. 2008. How to test for multiple realization. *Philosophy of Science* 75 (5):514–525.

Shettleworth, S. J. 2010. *Cognition, Evolution, and Behavior*. Oxford: Oxford University Press.

Short, T. L. 2007. *Peirce's Theory of Signs*. Cambridge: Cambridge University Press.

Shultz, T. R. 2003. *Computational Developmental Psychology*. Cambridge, MA: MIT Press.

Siegelmann, H. 1994. Analog computation via neural networks. *Theoretical Computer Science* 131 (2):331–360.

Simon, H. A. 1962. The architecture of complexity. *Proceedings of the American Philosophical Society* 106 (6):467–482.

Simon, H. A. 1979a. Information processing models of cognition. *Annual Review of Psychology* 30:363–396.

Simon, H. A. 1979b. *Models of Thought*. New Haven: Yale University Press.

Simon, H. A. 1993. The human mind: The symbolic level. *Proceedings of the American Philosophical Society* 137 (4):638–647.

Simon, H. A. 1996. *The Sciences of the Artificial*. Cambridge, MA: MIT Press.

Simon, H. A., and A. Newell. 1956. Models: Their uses and limitations. In *The State of the Social Sciences*, 66–83, ed. L. D. White. Chicago: University of Chicago Press.

Simon, H. A., and A. Newell. 1958. Heuristic problem solving: The next advance in operations research. *Operations Research* 6 (1):1–10.

Singh, R., and C. Eliasmith. 2006. Higher-dimensional neurons explain the tuning and dynamics of working memory cells. *Journal of Neuroscience: The Official Journal of the Society for Neuroscience* 26 (14): 3667–3678.

Sloman, A. 1996. Beyond Turing equivalence. In *Machines and Thought: The Legacy of Alan Turing*, 1:179–219, ed. P. Millican. New York: Oxford University Press.

Sloman, A. 2010. What's information, for an organism or intelligent machine? How can a machine or organism mean? In *Information and Computation*, ed. G. Dodig-Crnkovic and M. Burgin. Singapore: World Scientific.

Sloman, A. 2011. Comments on "The Emulating Interview . . . with Rick Grush." *Avant* 2 (2):35–44.

Smolin, L. 1997. *The Life of the Cosmos*. New York: Oxford University Press.

Sperling, G. 1960. The information available in brief visual presentations. *Psychological Monographs* 74 (11):1–29.

Steels, L. 2008. The symbol grounding problem has been solved, so what's next? In *Symbols and Embodiment: Debates on Meaning and Cognition*, 223–244, ed. M. de Vega, A. M. Glenberg, and A. C. Graesser. Oxford: Oxford University Press.

Sterelny, K. 1990. *The Representational Theory of Mind: An Introduction*. Oxford: Blackwell.

Sternberg, S. 1969. The discovery of processing stages: Extensions of Donders' method. *Acta Psychologica* 30:276–315.

Sun, R. 2002. *Duality of the Mind: A Bottom-Up Approach toward Cognition*. Mahwah, NJ: Lawrence Erlbaum.

Sun, R., L. A. Coward, and M. J. Zenzen. 2005. On levels of cognitive modeling. *Philosophical Psychology* 18 (5):613–637.

Suppes, P. 1962. Models of data. In *Logic, Methodology, and Philosophy of Science: Proceedings of the 1960 International Congress*, ed. E. Nagel, P. Suppes, and A. Tarski, 252–261. Stanford: Stanford University Press.

Taddeo, M., and L. Floridi. 2005. Solving the symbol grounding problem: A critical review of fifteen years of research. *Journal of Experimental & Theoretical Artificial Intelligence* 17 (4):419–445.

Taddeo, M., and L. Floridi. 2007. A praxical solution of the symbol grounding problem. *Minds and Machines* 17 (4):369–389.

Thelen, E., G. Schöner, C. Scheier, and L. B. Smith. 2001. The dynamics of embodiment: A field theory of infant perseverative reaching. *Behavioral and Brain Sciences* 24 (1):1–34, discussion 34–86.

Tolman, E. C. 1939. Prediction of vicarious trial and error by means of the schematic sowbug. *Psychological Review* 46 (4):318–336.

Tolman, E. C. 1948. Cognitive maps in rats and men. *Psychological Review* 55 (4):189–208.

Touretzky, D. S., and D. A. Pomerleau. 1994. Reconstructing physical symbol systems. *Cognitive Science* 18 (2):345–353.

Trout, J. D. 2008. Seduction without cause: Uncovering explanatory neurophilia. *Trends in Cognitive Sciences* 12 (8):281–282.

Trullier, O., S. I. Wiener, A. Berthoz, and J. A. Meyer. 1997. Biologically based artificial navigation systems: review and prospects. *Progress in Neurobiology* 51 (5):483–544.

Turing, A. 1950. Computing machinery and intelligence. *Mind* 59 (236):433–460.

van Gelder, T. 1995. What might cognition be, if not computation? *Journal of Philosophy* 91:345–381.

van Rooij, I. 2008. The tractable cognition thesis. *Cognitive Science* 32 (6):939–984.

Vera, A. H., and H. A. Simon. 1993. Situated action: A symbolic interpretation. *Cognitive Science* 17:7–48.

Walmsley, J. 2008. Explanation in dynamical cognitive science. *Minds and Machines* 18 (3):331–348.

Walter, W. G. 1950. An imitation of life. *Scientific American* 182 (5):42–45.

Wason, P. C. 1966. Reasoning. In *New Horizons in Psychology*, ed. B. M. Foss, 135–151. Harmondsworth: Penguin.

Watkins, M. J. 1990. Mediationism and the obfuscation of memory. *American Psychologist* 45 (3):328–335.

Webb, B. 1995. Using robots to model animals: A cricket test. *Robotics and Autonomous Systems* 16 (2–4):117–134.

Webb, B. 2000. What does robotics offer animal behaviour? *Animal Behaviour* 60 (5):545–558.

Webb, B. 2001. Can robots make good models of biological behaviour? *Behavioral and Brain Sciences* 24 (6):1033–1050, discussion 1050–1094.

Webb, B. 2004. Neural mechanisms for prediction: Do insects have forward models? *Trends in Neurosciences* 27 (5):278–282.

Webb, B. 2008. Using robots to understand animal behavior. *Advances in the Study of Behavior* 38:1–58.

Webb, B. 2009. Animals versus animats, or why not model the real iguana? *Adaptive Behavior* 17 (4):269–286.

Webb, B., and T. R. Consi, eds. 2001. *Biorobotics: Methods and Applications*. Cambridge, MA: MIT Press.

Webb, B., and T. Scutt. 2000. A simple latency-dependent spiking-neuron model of cricket phonotaxis. *Biological Cybernetics* 82 (3):247–269.

Webb, B., and J. Wessnitzer. 2009. Perception for action in insects. In *Spatial Temporal Patterns for Action-Oriented Perception in Roving Robots*, ed. P. Arena and L. Patanè, 3–42. Berlin: Springer Verlag.

Weisberg, D. S., F. C. Keil, J. Goodstein, E. Rawson, and J. R. Gray. 2008. The seductive allure of neuroscience explanations. *Journal of Cognitive Neuroscience* 20 (3): 470–477.

Weisberg, M. 2007. Three kinds of idealization. *Journal of Philosophy* 104 (12): 639–659.

Weizenbaum, J. 1976. *Computer Power and Human Reason: From Judgment to Calculation*. San Francisco: W.H. Freeman.

Wheeler, M. 2005. *Reconstructing the Cognitive World*. Cambridge, MA: MIT Press.

Wickelgren, W. A. 1969. Context-sensitive coding, associative memory, and serial order in (speech) behavior. *Psychological Review* 76 (1):1–15.

Wilson, R. A. 2004. *Boundaries of the Mind: The Individual in the Fragile Sciences - Cognition*. Cambridge: Cambridge University Press.

Wimsatt, W. C. 1997. Aggregativity: Reductive heuristics for finding emergence. *Philosophy of Science* 64:372–384.

Wimsatt, W. C. 2002. Functional organization, analogy, and inference. In *Functions: New Essays in the Philosophy of Psychology and Biology*, ed. A. Ariew, R. Cummins, and M. Perlman, 173–221. Oxford: Oxford University Press.

Wimsatt, W. C. 2007. *Re-Engineering Philosophy for Limited Beings: Piecewise Approximations to Reality*. Cambridge, MA: Harvard University Press.

Wood, W., and D. T. Neal. 2007. A new look at habits and the habit-goal interface. *Psychological Review* 114 (4):843–863.

Woodward, J. 2002. There is no such thing as a ceteris paribus law. *Erkenntnis* 57 (3):303–328.

Woodward, J. 2003. *Making Things Happen*. Oxford: Oxford University Press.

Woodward, J. 2008. Cause and explanation in psychiatry. In *Philosophical Issues in Psychiatry: Explanation, Phenomenology, and Nosology*, ed. K. S. Kendler and J. Parnas. Baltimore: Johns Hopkins University Press.

Wright, L. 1973. Functions. *Philosophical Review* 82 (2):139–168.

Zednik, C. 2011. The nature of dynamical explanation. *Philosophy of Science* 78 (2):238–263.

Zeigler, B. P. 1974. A conceptual basis for modelling and simulation. *International Journal of General Systems* 1 (4):213–228.

Index